The Great Chase: A History of World Whaling

Daniel Francis is a freelance historical researcher and writer, whose previous books include *Arctic Chase: A History of Whaling in Canada's North*, *Discovery of the North: The Exploration of Canada's Arctic* and *Battle for the West*. He has a B.A. in English from the University of British Columbia and an M.A. in Canadian Studies from Carleton University. He has also worked as a newspaper reporter and, from 1984 to 1987, was editorial director of the Horizon Canada Project. Daniel Francis is married with two children and now lives in Vancouver.

The Great Chase

A HISTORY OF WORLD WHALING

Daniel Francis

Penguin Books

PENGUIN BOOKS
Published by the Penguin Group
Penguin Books Canada Ltd, 10 Alcorn Avenue, Toronto, Ontario,
Canada M4V 1E4
Penguin Books Ltd, 27 Wrights Lane, London W8 5TZ, England
Penguin Books USA Inc., 375 Hudson Street, New York, New York
10014, U.S.A.
Penguin Books Australia Ltd, Ringwood, Victoria, Australia
Penguin Books (NZ) Ltd, 182-190 Wairau Road, Auckland 10,
New Zealand

Penguin Books Ltd, Registered Offices: Harmondsworth, Middlesex,
England

First published in Viking by Penguin Books Canada Limited, 1990

Published in Penguin Books, 1991

10 9 8 7 6 5 4 3 2 1

Manufactured in Canada

Canadian Cataloguing in Publication Data
Francis, Daniel
  The great chase

First published under title: A history of world whaling.
ISBN 0-14-011489-0

1. Whaling – History. I. Title. II. Title: A history of world whaling.

SH383.F73 1991   639′.28   C88-094770-5

Acknowledgements

Every effort has been made to acknowledge all sources of material used in this book. The publishers would be grateful if any errors or omissions were pointed out, so that they might be corrected.

F.D. Ommanney, *Lost Leviathan*
Copyright © F.D. Ommanney, 1971.
Reproduced by permission of Curtis Brown Ltd., London.

F.P. Schmitt et al., *Thomas Welcomes Roys: America's Pioneer of Modern Whaling*
University Press of Virginia, 1980.
The Mariners' Museum, Newport News, Virginia, U.S.A.
Reproduced by permission.

Edouard Stackpole, *Whales and Destiny*
Copyright © University of Massachusetts Press, 1972.
Reproduced by permission.

There is something extremely painful in the destruction of a whale . . . yet the object of the adventure, the value of the prize, the joy of the capture, cannot be sacrificed to feelings of compassion.

– Captain William Scoresby, Jr., 1820

The Great Chase

Table of Contents

PART TWO
WHALING IN THE MODERN AGE   171

Preface

This book began one summer afternoon in the early 1970s out on the rolling back of the St Lawrence River. Whale watching had not yet become the multi-million dollar industry it is today. There were perhaps two dozen of us, attracted by a small advertisement in a Montreal newspaper offering a two-day whale-spotting cruise on the river, sponsored by the Montreal Zoological Society and led by the society's director, Gerald Iles.

As our boat, a battered fisheries vessel borrowed for the expedition, crossed the river toward the small village of Tadoussac, we peered intently into the distance searching for the whales. The small boy in me wakened and I stationed myself right at the bow of the boat, determined to be the first to see one. Where we were, the St Lawrence is almost twenty miles across. As we approached the halfway point, I could see a flurry of whitecaps ahead, and wondered to myself why the water was suddenly so rough on such a calm afternoon. It was not until we drew near that I realized I was getting my first look at a whale in its natural habitat.

They were white whales, belugas, a large herd of them.

Their smooth backs rose and fell rhythmically in the stream; they looked like a splatter of white stones someone was throwing into the water. For several hours our boat followed them as they passed like a squall up the river. There was an innocence to the afternoon. For most of us, it was our introduction to the startling beauty and grace of whales in water. They seemed a revelation. Their white bodies plunged through the blue-green element, glowing in the sunshine like newly-fallen snow.

No one yet suspected that the river was becoming a poisonous sewer, that the whales would soon be so contaminated by industrial waste that they would be considered the most polluted mammals on earth.

Later in the day I stood at the rail of the boat, staring down idly into the water, resting my eyes for a moment from the intense glare of sunlight on the river. Quite suddenly a shape appeared in the depths, rising speedily toward me. It was a black whale, a sei, much larger than the beluga we'd been trailing. As it broke the surface directly below me, our eyes met and we looked at one another. Well, I certainly looked at it; perhaps the whale had other things on its mind. The moment passed. The animal dove back beneath the boat. Reclaiming my wits, I shouted, "Whale!" But the others had seen it emerge on the other side, and already they were busy with their cameras and binoculars.

That first encounter with the whales almost twenty years ago kindled a fascination with the animals and the sad history of their treatment at the hands of humans. I wanted to read about whaling in Canada, but when I went in search of books I discovered there were none. So, I did what any writer would do, I wrote my own. Titled *Arctic Chase*, it is an account of the American and British whaling industries in the Canadian Arctic.

But one book did not satisfy my appetite for the subject. Arctic whaling was just one brief chapter of a story that went back hundreds of years and spanned most of the world's oceans. I began travelling to some of the places connected with the whales and their hunters: to the tiny village of Telegraph Cove on Vancouver Island's northeast coast, embarkation point for a delightful day of cruising in Johnstone Strait, summer home to a shifting population of killer whales; to Coal Harbour on the far side of Vancouver Island, where a weathered jawbone of a huge blue whale stands as a monument to the period from 1948 to 1967 when the village was the site of a thriving whaling station; to New Bedford, Massachusetts, once the wealthiest town in the

United States thanks to whaling, where I stood in the chapel which Herman Melville immortalized in the pages of *Moby Dick* and read the marble slabs which hang on the walls like tombstones, each telling the tragic story of a life lost in pursuit of leviathan; and out to the island of Nantucket, the cradle of the American whaling industry, where I stayed in a house that once belonged to Archaelus Hammond, the first commercial whaler to harpoon a whale in the Pacific Ocean, and listened to Orrin Macy, an eleventh-generation islander, in the cluttered museum on Broad Street, tell stories of Nantucket seamen and their adventures on the whaling grounds on the other side of the world.

As far as possible I have tried to tell the story of whaling in the words of the whalers themselves. In the nineteenth century and earlier, it was an exotic, dangerous occupation, and not a few sailors turned author with accounts of their voyages to distant oceans. Some, like the young Englishman Frederick Bennett, wrote to alert the public to the gruesome conditions aboard the wooden whaling ships. Others, like the Nova Scotian Benjamin Doane, wrote to tell a younger generation what it was like to travel the world in the age of wind and sail. Others, like the not-yet-famous Herman Melville, wrote pot-boilers to feed the public's fascination with the South Pacific islands. And still others, like the surgeon Dr Charles Edward Smith, locked in the Arctic ice, wrote out of despair of ever seeing the green shores of his homeland again. All of their stories, and many others, are part of the history of whaling.

Like any voyager who wanders so far, I have incurred many debts: to librarians and archivists who took the trouble to hunt down obscure volumes; to other historians and scientists, without whose books on whales and whaling my own work of synthesis would not have been possible; to my wife and children who patiently endured my enthusiasms. In particular I would like to thank my friends Stephen Osborne, who read part of the manuscript and tried to help me find a voice, and Randall Reeves, who encouraged the project from the beginning, was generous with books and advice, and in the end made many useful comments about the manuscript. They have done their best to help; the errors and opinions that remain are solely my own.

—Daniel Francis, Vancouver, October, 1989

Introduction

From the cave paintings of Neolithic Man to the story of Jonah to Moby Dick, whales have long had a strong, even mystical, hold on the human imagination. In the Bible, Leviathan represents evil and death. During the Renaissance, drift whales cast up on the coast of Europe were interpreted as portents of the future. In our own day, recorded songs of the humpback whale have been launched deep into space, accompanying human greetings to extraterrestrial worlds. The concern, even moral indignation, that greets the death of one of these creatures far exceeds our solicitude for any other animal.

Our culture has at least two theories about the origins of the whale. One is biblical. In the story of Genesis, the whale is singled out for special mention. "And God created great whales, and every living creature that moveth, which the waters brought forth abundantly, after their kind, and every winged fowl after his kind: and God saw that it was good. And God blessed them, saying, Be fruitful and multiply, and fill the waters in the sea, and let fowl multiply in the earth. And the evening and the morning were the fifth day." (Genesis 1: 21–23)

The other theory is scientific. It traces the origins of the

1

whale back about sixty million years to a furry, four-legged land animal that foraged for food at the edge of lagoons and estuaries. Following its prey into deeper water, this ancient mammal returned permanently to the sea, shedding its fur, growing fins and evolving into the more than seventy species of whales, dolphins and porpoises known to man.

Over the years whalers hunted many of these species, but the industry was sustained always by the larger animals, the so-called great whales. The first victims were the slow-swimming right whales of the North Atlantic, followed by their near relatives, the Arctic-dwelling bowhead. After depleting the inshore populations of these species, whalers moved out into deeper water to take on the sperm whale. This was the heyday of American whaling, described by Herman Melville in his classic novel, *Moby Dick*. At the same time a brief but devastating hunt for the gray whale almost exterminated that species. In our own century the whalers waged total war on the whales of the southern hemisphere—blues, fins, humpbacks and seis. Today the only baleen species that survive in healthy numbers in the southern ocean are the minke and the Bryde's whale, the smallest of the great whales.

This is a book about whaling, not whales. However, to understand the industry it is helpful to know something about the animal. The great whale is an improbable beast. Like other mammals, including man, it is warm-blooded, breathes air, gives live birth and suckles its young with milk. Yet it lives its entire life in the water, and as a result was long thought to be a species of gigantic fish. Melville's shorthand definition of a whale was "a spouting fish with a horizontal tail," and whalers commonly referred to their enterprise as the "fishery."

The animal's dimensions alone defy belief. The largest creature known to have lived on earth, the blue whale grows as long as a railway boxcar and weighs as much as twenty-five elephants. At birth, a calf already weighs several tons, and proceeds to gain weight at the incredible rate of eight pounds an hour. Everything about the whale is immense, from its automobile-sized heart, to its hundred-pound testicles, to its aorta the size of a water main. There is enough blood in a large whale to fill seven thousand milk bottles. Its tongue alone would need a flat-bed truck to carry it. The tail flukes are broad enough to fly an airplane. And so the surprising statistics accumulate.

What possessed the whalers to take on such a giant? The answer lies in the thick layer of blubber that envelops the animal. Much like a blanket, blubber insulates the whale and keeps its body temperature constant, no matter what the temperature of the water through which it is swimming. It is also a storehouse of energy that the animal draws on when food is scarce. The thickness of the blubber varies anywhere from two inches to two feet depending on the season of the year and the species of whale. It is composed of white, fibrous, fatty material honey-combed with large cells filled with oil. When exposed to high heat, the oil separates from the fibres, producing the liquid gold that was the whole purpose of the whaling enterprise. Whale oil was an excellent lighting fuel in the days before gas and electricity, a high-quality lubricant, a cleanser in the woollen textile industry, and a basic ingredient in the manufacture of soap and, in this century, margarine.

The great whales belong to two distinct suborders of the order *Cetacea*. One, the *Odontoceti*, has teeth. For the whalers, the most important of the *Odontoceti* was the sperm whale, hunted for its oil and spermaceti, the waxy substance found in the animal's head. The second suborder is the *Mysticeti*, the "moustached whales." Most of the great whales are members of this group. Instead of teeth, the *Mysticeti* have strips of baleen, a horny substance much like human fingernails, growing from their upper gums. The inner edge of each baleen piece, also known as whalebone, is frayed and bristly. When the whale opens its mouth, the baleen descends like a curtain across the opening. Taking a mouthful of seawater, the animal forces it back out through the baleen. In the process, food is retained in the bristles, then swallowed.

Individual baleen plates grow to a length exceeding twelve feet depending on the species. This flexible substance was in great demand before the invention of spring steel and plastic. Indeed, for a period in the nineteenth century when the price of whale oil plummeted, the whaling industry survived only because of the continued demand for baleen.

A third product of the hunt, whale meat, has never been in great demand in Europe or America. For many years the whalers threw it back into the sea. More recently, whalers have ground it up for fertilizer and animal feed. The exception are the Japanese, who made whale meat a staple of their diet, and the native inhabitants of Alaska and the Canadian Arctic. Incidentally, following accepted practice, I have used the term Inuit for natives of the Canadian Arctic and Eskimo when referring to Alaskan native inhabitants.

Whalers were not much interested in the complex and fascinating biology of the whale. Only one thing interested them—where the animals were to be found. Most whales are migratory animals, ranging vast distances in search of food, often returning to specific locations to give birth to their young. The gray whale, for instance, makes the round trip between the calving lagoons of Baja California and its feeding grounds in the Bering Sea each year like clockwork. Similarly, other species of baleen whales migrate seasonally along the coasts of Australia, Africa and South America. Even the nomadic, deep-water sperm whale tends to congregate in certain stretches of ocean at certain times of the year. Such regularity made it easy for the whalers to intercept their prey.

Whaling relies on one simple biological fact. Whales breathe air. Though they live most of their lives out of sight in the ocean's depths, they must return regularly to the surface to take a breath. And there the whalers wait, watching for the tell-tale spout.

A whale's blowholes (or blowhole in the case of toothed whales) are nostrils that have migrated to the top of the animal's head. The spout, sometimes rising twenty-five feet into the air, is not a fountain of water. It is a misty spire of vapour, mixed with oily mucus, that is exhaled in a mighty blast from the lungs. Whales are not at their most efficient on the surface. Their streamlined bodies move through water far more easily than through the turbulence of wind and waves. However, before they can return to their underwater world they must replenish their supply of oxygen, taking a series of breaths that become more rapid as the animal prepares to dive.

Baleen whales feed on fish and plankton that inhabit the upper layer of the ocean, but a predator like the sperm whale can dive over a mile into the black depths in search of food, remaining below for up to two hours. At such extreme depths, the pressure is great enough to crush most other animals. But sperm whales do not even get the bends. They are not simply holding their breath. Whales, in fact, have small lungs relative to the size of their massive bodies. Most of the oxygen that a sperm whale needs during a dive does not come from the lungs at all. Instead, it is stored in high concentrations in the blood and muscles. At the same time, the animal reduces its heart rate and limits the blood flow to parts of the body other than the heart and brain. Nitrogen, so dangerous to human divers when it forms embolisms in the blood stream and tissues, is neutralized by the whale's system.

However long the whale remains underwater, it must return to the surface eventually, which is when the hunt begins. Humans have been hunting whales for thousands of years, harpooning them on the open seas or herding them into shallow estuaries. For the most part, this early hunt was carried on by aboriginal peoples to supply their own needs for food and fuel. Large-scale commercial hunting began with the Basques of southern Europe who were chasing right whales in the Bay of Biscay by the eleventh century, perhaps even earlier. From Iberia the hunt spread north to the ice-choked waters of the Arctic, and west across the Atlantic to the shores of the New World. One after another the whaling grounds were discovered and the stocks of whales depleted.

At the end of the nineteenth century, the machinery of industrialism was applied to whaling, with devastating results. Equipped with exploding harpoons, speedy catcher boats and giant factory ships, whalers scoured the world in search of their prey. Modern-style whaling decimated the whales. Between the 1920s and the 1970s, more than two million were killed, and several species were reduced to the verge of extinction.

Since the 1970s the massacre has slowed. In 1982 members of the International Whaling Commission agreed to a moratorium on commercial whaling, a breathing space to assess the condition of the world's whales. The moratorium came into effect in 1986. Though hunters still kill a few hundred animals each year, ostensibly for scientific reasons, and in the case of the Inuit, for subsistence, the commercial hunt is on hold, at least until the IWC reassesses the moratorium in 1990.

Everyone now seems to agree that endangered whale species—the blue, the right and the bowhead, to name just three—should be spared further killing. But there is less agreement about other species—the minke is usually mentioned—which seem to be healthy enough to sustain a limited commercial hunt. It is possible that there is no compelling *scientific* reason not to allow a controlled harvest of these whales. For those who endorse whaling, the whale is not a special case, but rather a marine resource that has to be managed like any other. If stocks are large enough to sustain a hunt, they say, then let the hunt begin.

But environmentalists argue that no commercial whaling should be allowed. They feel it is pointless economically, since the animal produces nothing for which a substitute product cannot be found. Many also believe that the whale is a special animal of great intelligence which it is unethical to kill. In any case, the whale has become a potent symbol for the environmental movement. Environmentalists have hailed the suspension of the commercial hunt as a great victory, evidence that the world can be saved from its apparent downward spiral into ecological catastrophe. As a result, many people believe that no whales at all should be killed, regardless of the arguments of science. For these people, the fate of the whales portends the fate of the world, and a resumption of whaling would be an outrage.

My own view is that the whalers base their arguments as much on faith as on science. First of all, whales are migratory animals that spend most of their lives underwater. In other words, they are hard to count. The census takers have been wrong before, and the whales have suffered for it. At the annual meeting of the Whaling Commission in 1989, scientists presented data suggesting that the blue whale population in Antarctica may be one-tenth the size everyone had confidently believed. Such shocking news must make us suspicious of any claims that other species are thriving.

Second, the condition of the whales cannot be considered in isolation from the condition of their habitat, the world's oceans. All the evidence points to the steady deterioration in the quality of the oceans as man continues to pour pollutants into them like poison down a sink. Oil spills, an accident involving a nuclear submarine, chemical leaks—one can think of several scenarios that might suddenly and catastrophically alter the whales' environment. How can we say that any stock of whales is healthy when increasingly we are destroying the world in which they live?

Of course, these arguments do not take into account the political realities that will face members of the Whaling Commission when they come to make a decision about the moratorium. If some countries are determined to resume whaling in the face of world opinion, the Commission can do little about it. IWC delegates may well decide, as they have in the past, that it is better to allow a small hunt, and keep the whaling nations in the family, than it is to force the whalers out of the organization—where their activities would be totally unregulated.

Meanwhile, it is vital for us to understand the history of mankind's relationship with one of the most awesome animals on the planet.

Chronology of World Whaling

1100–Basque whalers are hunting whales from shore in the Bay of Biscay. The Basques are considered the first large-scale commercial whalers.

ca. 1400–Basque whalers begin extending their range out of the Bay of Biscay into the North Atlantic.

ca. 1550–Basque whalers have reached the New World and are spending summers at shore stations along the north shore of the Strait of Belle Isle in Labrador.

1607–Henry Hudson returns from a voyage in search of a Northeast Passage with a report that the waters near Spitsbergen are teeming with whales. Within a few years, British and Dutch whaleships are competing to control the Spitsbergen grounds.

1613–Dutch merchants combine to form the Noordsche Compagnie and win a monopoly over the northern whaling grounds. The Noordsche Compagnie enjoys its monopoly until 1642.

1617–Dutch whalers on Amsterdam Island erect the first buildings of their main shore station at Spitsbergen. This place, called Smeerenburg, grows into a thriving summer settlement until it is abandoned in the 1640s as the whaling begins to move offshore.

ca. 1650–Shore whaling begins on Long Island, New York. This represents the beginning of the American whaling industry.

ca. 1670–The Dutch whaling fleet is the largest and most productive in the world and dominates the industry.

1672–Residents of Nantucket kill a large whale that is trapped in the island's main harbour. This event is considered the beginning of Nantucket's whaling industry.

ca. 1712–A Nantucket whaling captain, blown out to sea in a gale, kills a sperm whale and inaugurates American deep-water whaling. Soon New England vessels are cruising on several hunting grounds in the Atlantic.

1720–Dutch whalers in increasing numbers are visiting Davis Strait to hunt bowhead whales.

1749–In an attempt to stimulate its whaling industry, the British government offers to pay a bounty of 40 shillings per ton to every whaleship weighing more than 200 tons. This subsidy encourages a rebirth of British whaling, and by the end of the century the British have replaced the Dutch as the most important Arctic whalers.

ca. 1750–The development of the spermaceti candle provides a huge new market for sperm whale products. About the same time American captains begin introducing tryworks onto their vessels, meaning that blubber from the whales can be boiled during a voyage and does not have to be brought back to shore for processing. This frees the whalers to make longer and longer voyages, especially into the South Atlantic.

1775–The American Revolution begins, and for the next decade the Yankee whaling industry is moribund.

1786–Nantucket whaling merchant William Rotch establishes a whaling colony at Dunkirk in France.

1789–In January the British ship, *Emilia*, becomes the first whaling vessel to enter the Pacific Ocean.

1791–A British vessel undertakes the first whaling voyage near Australia.

1813–American naval vessel attacks and destroys about 50 percent of the British Pacific whaling fleet.

1817–Two British whaleships make the first crossing of Baffin Bay to the "west water" near the mouth of Lancaster Sound, initiating whaling along the coast of Baffin Island.

1818–Discovery of the new "off-shore" grounds teeming with sperm whales out in the Pacific west of Peru.

1819–Whalers begin visiting the Hawaiian Islands.

1830–Nineteen British whaleships in Davis Strait destroyed by ice and storm, the worst season of destruction in the history of Arctic whaling.

1835–Discovery of the Northwest Grounds off the coast of what is now British Columbia, where right whales were particularly plentiful.

1845–The hunt for gray whales begins in the lagoons of Baja California. The hunt lasts through the 1860s and almost wipes out the population of grays.

1847–Nine hundred vessels are now taking part in whaling around the globe, the result of a long period of expansion in the industry since 1815. The British are still active in the Arctic; otherwise, this is the era of American dominance.

1848–The first whaleship sails through Bering Strait into the Arctic.

1851–Herman Melville publishes his classic whaling novel, *Moby Dick*.

–First shore whaling station established on Baffin Island by a party of American whalers.

1857–British whalers begin installing steam engines in their ships.

1859–Oil discovered in Pennsylvania; eventually this discovery leads to a drastic decline in the demand for whale oil.

1863–Svend Foyn builds his new, steam-driven schooner, the *Spes et Fides*, the prototype for the powerful, speedy catcher boats that will revolutionize the whale hunt.

1865–A New Bedford whaling captain invents the darting gun, a form of explosive harpoon.
    –During the American Civil War, the Confederate warship *Shenandoah* destroys 34 whaleships on a surprise raid into the western Arctic.

1868–Svend Foyn combines all the elements of his new whale-catching technique at his station in northern Norway. His innovations mark the dividing line between old-fashioned harpoon whaling and modern industrial whaling.

1871–Thirty-three American whaleships destroyed in the ice off Alaska.

1880–The maiden voyage of the first American whaleship powered by steam, the *Mary and Helen*.

1890–American whalers begin overwintering at Herschel Island in the western Arctic.

1892–The first whaling expedition to the Antarctic meets with disappointing results.

1904–Shore whaling begins at the island of South Georgia, headquarters for the Antarctic hunt.

1907–The price of whalebone has fallen so low that the American bowhead whaling industry is crippled and never recovers.

1909–Through a process called hydrogenation, whale oil is solidified and purified so that it can be used in the production of better quality soaps and margarine. Once again, whale oil is in great demand.

1925–The first factory ship with a stern slipway arrives on the Antarctic grounds. These huge vessels give a new level of efficiency to the hunt.

1929–Norway passes the first legislation to control whaling on the high seas. The Norwegian Whaling Act provides the model for much of the whaling regulation that follows.

1931–A record 29,400 blue whales are killed by whalers in the Antarctic.
    –The Geneva Convention for the Regulation of Whaling is the first attempt by the world body to control the whale kill.

1937–International Agreement for the Regulation of Whaling is signed in London. An improvement on the Geneva accord, it still stops short of effectively protecting the whales.

1944–A conference of whaling nations in London introduces the Blue Whale Unit and sets a limit on the number of whales that may be killed each year.

1946–The International Whaling Commission is created, but it fails to halt the decline in the population of the world's whales.

1972–Mexico declares Scammon's Lagoon (Laguna Ojo de Liebre) the world's first whale refuge.
    –United Nations Conference on the Human Environment calls for a 10-year moratorium on commercial whaling. Members of the IWC reject the idea.

1974–The IWC introduces its New Management Procedure, which is supposed to manage the whale kill scientifically.

1975–The environmental group, Greenpeace, begins its campaign against commercial whaling.

1979–Paul Watson's vessel, the *Sea Shepherd*, rams the pirate whaler *Sierra* off Portugal. Later the *Sea Shepherd* is scuttled by Watson, and the *Sierra* is blown up.

1982–The IWC votes to accept a 10-year moratorium on whaling. Some whaling nations refuse to abide by the decision; others continue to whale for so-called scientific purposes.

1986–"Eco-guerillas" sabotage a whaling station in Iceland, sparking a debate about the limits of environmental protest.

1989–Scientists report to the IWC that the population of blue whales in the Antarctic may be only ten percent of previous estimates.

1990–The IWC debates whether to continue the moratorium on commercial whaling.

PART ONE

OLD-STYLE WHALING

FIRST WHALERS

The Basques were then the only people who understood whaling.[1]

—Captain Jonas Poole, 1610

Summer departs the treeless shores of Labrador all of a sudden. One day the rocks are warm with September sunshine; the next they are glazed with frost and the waves that lash the beach throw up a freezing spray. Summer visitors must understand the harsh etiquette of this northern coast: when it is safe for them to stay, when it is best to go. The wrong decision may be fatal.

In the fall of 1565, Joanes de Portu, a whaling captain from the Basque country of Spain, made the wrong decision. He lingered too long at his anchorage in the harbour of Red Bay in the Strait of Belle Isle at the eastern edge of America. While he prepared his galleon, the *San Juan*, for the three-thousand-mile voyage back across the Atlantic to Spain, a violent storm swept down out of the interior. Black clouds tumbled over the surrounding headlands, driven by howling winds that whipped the surface of the bay into a frenzy. Sheets of rain lashed the backs of the sailors as they hurried to secure the vessel.

As the wind increased, the *San Juan* broke loose from its anchor line and drifted wildly toward shore. The sailors could do nothing now to save their vessel. It grounded stern first. Pounded again and again against the rocky bottom, its oak hull broke open and the ship foundered about thirty yards from land, listing over on its starboard side as it filled with water.

When the gale blew the *San Juan* ashore, the ship carried as many as a thousand casks of whale oil. In 1980s money, the cargo was worth between $4 and $6 million, a fabulous sum that shows the value of the Basque "fishery" in America. Captain de Portu did not intend to give it up without a struggle. His men salvaged the sails and rigging, some provisions, and as many casks as they could before hitching a ride back to Spain aboard another whaler. For at least a year the superstructure of the wreck remained visible above the surface. The next summer de Portu retrieved more of the precious oil casks. Then the effects of ice and storm shattered the hulk. Its broken planks and timbers sank into the silt at the bottom of the bay, and the *San Juan* was forgotten.[2]

Four hundred years after Joanes de Portu recovered his last oil barrel from the sunken wreck, the *San Juan* resurfaced. Mrs Selma Barkham, a researcher working for the Canadian government in the Spanish archives, came across several documents describing the fate of the Basque whaleship. Using her information, archaeologists located the wreck at Red Bay, as well as several other Basque vessels and the remains of buildings that once made this remote harbour one of the busiest ports in America.

In 1978, when divers first visited the wreck of the *San Juan*, it was a tangled mass of timbers, barrel staves and ballast stones, half-buried in the silt. Over the years the weight of the ice had split the hull wide open and flattened it onto the bottom of the bay. There was no way to lift the vessel intact to the surface; indeed, there was no vessel left intact. Instead the research team dismantled the remains of the smashed hull and brought it up piece by piece for study. Because they were expected to deteriorate after long exposure to the elements, timbers were resubmerged at the site of the wreck and buried in a huge pit, where they remain today.

The discovery of the *San Juan*, along with other evidence of Basque activity on the Labrador coast, reveals a well-organized whaling industry active in the New World not long after Columbus. Some people even claim that the Basques preceded Columbus to the shores of America by more than one hundred years, though there is no hard evidence to prove that was so.

Whether or not they crossed the Atlantic so soon, the Basques certainly were hunting whales out in the Bay of Biscay as early as the eleventh century. Eventually they rode their sailing galleons north to the coasts of Ireland and Iceland, then west across the Atlantic to *Terranova*, the "new land." They were Europe's first commercial whalers, and the techniques they pioneered were emulated by every other whaling nation.

I

The Basques are a unique people. Their exact origins are obscured by the mists of time, but they are among the oldest populations in Europe. For thousands of years, perhaps since Neolithic times (ca. 4000 B.C.), they have inhabited roughly the same corner of the western Pyrenees, where modern Spain and France touch. Originally pastoralists who tended their animal flocks in high mountain pastures, they had the misfortune to occupy a convenient access route into the Iberian Peninsula. From earliest times successive waves of invaders swept back and forth across their rugged homeland. But the Basques were tenacious people who were neither absorbed nor evicted by these trespassers. In time, they created their own kingdom, the Kingdom of Navarre, which early in the eleventh century expanded to include all Basque-speaking people on either side of the Pyrenees.[3]

By the eleventh century, as well, Basques inhabiting the tiny fishing villages of the Biscay coast were venturing farther out to sea and gaining a reputation among seafaring nations not only as fishermen but as shipbuilders, merchant seamen and, lastly, as whalers. No one knows for sure when the Basques began pursuing the whale; the earliest records date from the middle of the eleventh century. They may have learned from their northern neighbours, the Normans and the Vikings, who captured whales by herding them into shallows or narrow bays. However, the smoother shores of the Bay of Biscay have no such convenient geographical "traps," so when the Basques took up the hunt they developed their own techniques for chasing the animals from boats in open water. It was this adaptation of technique to the requirements of open-water whaling that earned for the Basques the reputation as the world's first modern whalers.[4]

Each fall the Bay of Biscay welcomed an annual migration of black whales (*Eubalaena glacialis*) heading south from their

Monsters of the deep. A detail from a 1546 map by Pierre Descelliers is the earliest depiction of European whalers at work.

summer habitat in the Norwegian Sea between Norway and Iceland. The Norwegians called these animals *nord-kapers*; the French knew them as *baleines des Sardes*; and the Basques called them *sardako*, "whales living in groups." But today they are known as right whales. Slow-swimming, docile in temperament and, most importantly, buoyant after death, they were the "right" whales for hunters to pursue.

These lounging giants, up to sixty feet in length, entered the bay in October and remained until March. During this period watchers climbed into lookout towers and lighthouses along the coast to keep an eye out for whales passing close to shore. When he sighted a spout, the watcher alerted the village by burning some straw, beating a drum, ringing a bell or waving a flag. At this signal, hunters rushed from their houses down to the beach where the boats sat ready for launching. If the shoreline was steep, boats were held in place on the slope by a rope attached to a capstan. Once they were on board, the rowers released the capstan and the boat slid downhill into the water with a great splash.

A harpooner commanded each boat, standing in the bow shouting directions back to the helmsman. When he arrived within striking distance, he thrust the V-shaped harpoon with as much force as he could into the back of the whale. This blow was not intended to kill the animal, simply to wound and stampede it. A rope attached the harpoon to a large, dried gourd that was tossed into the water to act as a drag and buoy. Startled and hurt, the whale surged away, but by keeping the buoy in sight the boats could follow, planting more harpoons when they could get close enough. Eventually the animal grew too tired to continue the chase. It lay exhausted on the surface, where one boat was able to draw close enough to administer the *coup de grâce*. Taking up a long, spear-like lance, the harpooner plunged it into the whale, churning it up and down in search of the heart or lungs. When the dying animal began to spout blood, the hunters knew they had pierced a vital organ. In a short time the whale went into its death flurry, then expired.

Basque whalers used two types of boats; a small, manoeuvrable craft for chasing the animals and a larger boat with a crew of ten that arrived after the kill to help tow the huge carcass back to shore. When they arrived back at the village with their catch, whalers waited for the high tide, then landed the beast as far up the beach as possible. From there they hauled it with winch and cable above the high-water mark and began removing the blubber in large slabs with razor-sharp flensing knives. These large pieces were cut up into smaller ones and taken to nearby boiling houses. There the blubber was boiled in huge cauldrons to separate the oil from the skin and tissue. The smell produced by the boiling oil and burning skin was so offensive that some villages refused to permit this process to be carried out in their vicinity.

The Basques found a market for almost every part of a whale carcass. Oil was as crucially important to Europeans in the Middle Ages and the centuries that followed as it is today, and whales were a major source. It was used to fuel the lamps in churches, public buildings and, much later, streets. It also served as a lubricant for machinery, to clean wool and soften leather, and in the preparation of soap, medicines and paint. As well as oil, a right whale has between 220 and 260 strips of baleen, also known as whalebone, in its mouth. They were dried, washed and moulded into a vast array of useful products, including umbrellas, fans, snuffboxes, chair springs, the bristles in hairbrushes, walking sticks, skirt hoops and corset stays. In the Basque coastal villages bones ended up as fence posts; skin was fashioned into belts, boots and bellropes. Unrendered blubber found popularity as a condiment, and whalers reserved the huge tongue of the animal as a special delicacy for town dignitaries. The bulk of the meat sold at market where it was considered a kind of fish and therefore suitable to be eaten in Catholic countries during the 166 days of the year that meat-eating was prohibited by the Church. Even the animal's excrement found a use dyeing fabrics red. So valuable were the products of a whale that a single kill in a season was enough to make the hunt profitable for one of the dozens of fishing villages strung around the edge of the bay.

II

By the beginning of the fifteenth century the Basques were extending their whaling voyages beyond the familiar confines of the Gulf of Gascony, the inner region of the Bay of Biscay. Looking for fresh whaling grounds, they sailed west along the "roof" of Spain and north up the Atlantic coast of Europe as far as Iceland, and even Greenland. It is unclear what drove the Basques away from local shores. Perhaps the number of whales was declining. Each winter shore whalers managed to kill only about a hundred animals, not a large number by modern standards but possibly enough to deplete the herds of right whales that visited the Bay of Biscay. More likely, some of the more ambitious Basque mariners recognized that by extending the range of their voyages they could extend the length of the hunting season and increase the size of their catch. No longer would they have to wait for the onset of autumn to bring the whales south to their shores; instead, the hunters would go to the whales.

The east coast of Canada has attracted European fishermen at least since the Italian mariner John Cabot returned from his voyage to Newfoundland in 1497 with stories of teeming stocks of cod that almost leapt into his boat. By the 1540s Basque whalers had joined the fleet of fishing vessels scurrying back and forth across the North Atlantic. Their interest centred on the bleak southern shore of Labrador. French explorer Jacques Cartier dismissed this barren stretch of rocky coast as "the land God gave to Cain" when he cruised along it in 1534. Icebound for much of the year, unfit for farming and inhabited by unknown native peoples, it offered no attractions for would-be colonizers. But the Basques found it far from unprofitable. They discovered that the narrow Strait of Belle Isle between Labrador and the northern tip of Newfoundland formed a natural funnel through which whales migrated into the Gulf of St Lawrence from the Atlantic. The strait is a unique meeting place where cold ocean currents from the north merge with warmer water rich in nutrients, from the south. As a result, tiny marine plants and animals, on which whales feed, grow in abundance. For the most part the Basques found right whales on this ground, the same species that they hunted closer to home in the Bay of Biscay. However, recent research indicates that bowhead whales, a species named for its immense, bowed head, were also present in the strait in this era and would certainly have been taken as well. The Basques called this area *Granbaya*, Grand Bay, and for most of the sixteenth century harbours along the north shore of the strait provided a summer base for hundreds of seamen who ventured three thousand miles from their homeland to pursue the whale.[5]

Red Bay was one of these harbours. Today it is a small fishing village of three hundred people, a scattering of white

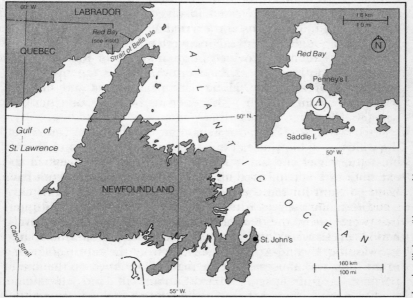

Red Bay, Labrador. The wreck of the San Juan lies at 'A'.

23

buildings along the bald shore at the end of a gravel road. The earliest Breton fishermen named it *Les Buttes*, and this reference to the red rock bluffs that stretch along the coast is retained in the present name. The road runs as far as Blanc Sablon, another fishing village about seventy-five miles away. The only way to reach this isolated spot is by plane or, like the ancient Basques, by boat.

The bay is roughly circular in shape; from the air it looks like a balloon that is partially deflated. The narrow entrance to the inner basin of the bay is fronted by a long island that rises to low hills at either end and is appropriately named Saddle Island. It was the shelter offered by this natural barrier that attracted Basque whalers to Red Bay. Nestling their ships in behind the island, they used its protected landward side as a site for their boiling house and workshops. Evidence of all these activities has been found by archaeologists, along with the shattered remains of the *San Juan*.

The slopes of the western Pyrenees were covered with forests of oak trees which Basque shipbuilders used to make the small, wooden caravels and larger galleons that brought the whalers to Labrador. At about three hundred tons' burden, the *San Juan* was among the smaller vessels involved in the whaling; others were seven hundred tons and carried a crew of up to 130 men and boys. Though cod and salmon were plentiful near the coastal harbours, victuallers provided large amounts of foodstuffs for the eight-month voyages. Each crew member ate up to two pounds of bread a day, along with peas, beans and slices of bacon, all washed down with generous helpings of cider and wine.

The voyage to Labrador began one day in May or June when the winds were blowing fair and the priest had been on board to bless the vessel and say a special mass for the success of the expedition. The trip lasted up to two months and could not have been pleasant for members of the crew. Living conditions were spartan. Sailors slept on the hard decks or on straw palliasses that were filthy and crawling with vermin before the voyage ended. Fresh water quickly spoiled, and none could be spared for washing. By mid-Atlantic the smell from the refuse collecting in the bilge was nauseating. One mariner summed up the dreary particulars of life at sea: "a hard Cabbin, cold and salte Meate, broken sleepes, mouldy bread, dead beere, wet Cloathes, want of fire." For the members of a whaling crew, however, it could be worthwhile. When they hunted close to home in Biscayan waters, a boat's crew seldom shared more than a couple of whales in a season. In Labrador the stakes were considerably higher. A ship expected to take anywhere from a dozen to twenty animals a year, and one-third of the oil produced went to pay the crew. Certainly the average sailor on a Basque vessel was better paid than his counterpart on an American whaleship in the heyday of Yankee whaling four hundred years later.

Navigation in the age of Basque whaling relied as much on intuition as science. Pilots calculated their latitude by the stars with the aid of an astrolabe and quadrant and plotted direction using a compass, but they had no way of determining their longitude or the ship's speed. Once in sight of land only the lead line and the sharp eyes of the lookout kept the vessel off the rocks. When in doubt the captain might consult his rutter, a book of sailing directions compiled from information passed on by other mariners, but in the narrow waters of the Strait of Belle Isle there was not much to guide him but his own knowledge of the coast picked up on earlier voyages.

Whalers wanted to be at Red Bay and the other harbours as soon as the ice disappeared in the spring. There was much work to do to prepare for the arrival of the whales. The barrels that carried the precious cargo of oil back to Europe were brought to Labrador in pieces to conserve space, and assembled on the spot by coopers. Most of the men lived on board their ships but the coopers moved onshore and their dwellings and workshops were erected from timber cut and hewn locally. As well, the crew had to build a structure to house the furnaces for boiling oil. At Red Bay this shelter was located on the lee side of Saddle Island, looking into the entrance to the bay. It consisted of a stone wall, about thirty feet long, running parallel to the shore. Fireboxes were built into the wall on the water side. Huge copper cauldrons filled with boiling blubber sat on top of these furnaces. On the landward side of the wall a raised wooden platform allowed tenders to wheel up loads of blubber and to ladle out the oil when it was ready. The furnaces were roofed over with red clay tiles brought by the thousands from Europe. Many of these tiles still litter the ground around the harbours.

The hunt for whales in the Strait of Belle Isle differed little from the shore fishery that the Basques had been pursuing in the Bay of Biscay for hundreds of years. It began when a lookout in one of the watchtowers spotted the first spout of the season, announcing the arrival of the whales from the north. Hunters went out in open boats called *chalupas*, measuring about twenty-five feet in length, each carrying a crew of six rowers and a steersman. The harpoon was attached to a drag designed to tire the whale as it tried to make its escape. Boats followed along behind, waiting for exhaustion to pacify the giant beast so that it could be killed with the lance. The chase often took several hours and ended many miles from the harbour. Several of the *chalupas* tied on to the carcass, which was far too heavy for a single boat to tow. If night fell before the hunters returned, their comrades built signal fires ashore to lead them home.

As soon as the dead whale arrived in the harbour, flensers set to work peeling off the layer of thick blubber. This process sometimes took place on the beach near the boiling house, but it was difficult to turn a whale lying on dry land without quite sophisticated machinery, so whales often were stripped as they lay in the water beside the ship and the pieces of blubber were carried ashore by boat. The brick ovens stood almost six feet high. Tenders fed fuel through a door in the front. Each cauldron held from fifty to two hundred gallons of oil which, when it had boiled to the right quality, was ladled into vats of water for cooling. Being lighter than the water, the oil remained on the surface where it was skimmed off and poured into the wooden *barricas*. A boat towed these floating casks out to the ship where they were stored in the hold.

The *San Juan*'s cargo of a thousand barrels of oil when it sank was not unusual. Basque whaleships customarily returned home from Labrador in December with full holds. During the peak years of the 1560s and 1570s, Red Bay alone produced up to nine thousand *barricas* of oil each season. In a good year the coast as a whole produced more than a million gallons of the precious liquid.

III

Towards the end of the 1500s Basque whalers began to withdraw from the Labrador grounds. Warfare between nations in Europe, culminating in the defeat of the Spanish Armada in 1588, destroyed much of the Basque merchant fleet and ruined many investors. Whalers who remained in the business suffered from the damaging tax policies of the Spanish government and began to seek higher returns in the fur trade and the cod fishery. There is also some evidence that natives along the Labrador coast, particularly the Inuit, began to resent the presence of several hundred strangers on their land each year and grew increasingly hostile. There is also evidence that the population of whales on the coast might have been declining as a result of over-hunting. The Basques killed about three hundred animals a year and it was part of their strategy to take calves and mothers wherever possible, because they were easier to catch. If the whales were not disappearing, at least their numbers might have been so depleted as to make other, more plentiful, hunting grounds more attractive to Basque whalers.

For this combination of reasons, the Basques left the Strait of Belle Isle. The last recorded visit to Red Bay was in 1604; by 1620 the entire stretch of coast may have been abandoned. Harbours that had echoed with the industry of several hundred Basque seamen fell silent. The boiling houses began to deteriorate and collapse in piles of tile and rubble. Heaps of bleached bones lay on the shore, in the words of one visitor, "like overturned tree trunks one atop the other." In the whalers' cemetery at one end of Saddle Island lay the untended graves of Basque whalers who ended their lives buried in this rocky soil so far from home.

Some of the whalers moved their operations deeper into the Gulf of St Lawrence and continued whaling through the seventeenth century. But most of those who did not withdraw from the industry altogether switched their activities to a brand new whaling ground discovered in 1607 near Spitsbergen. The discovery of Spitsbergen, a frozen archipelago in the Arctic Ocean within six hundred miles of the North Pole, opened a new chapter in the history of world whaling, one in which the Dutch and the British would first of all learn from the Basques, then drive these pioneers from the business.

WHALE WARS

*Concerning the whale's valour, we do find that he is not
very courageous, according to his strength and bigness;
for if he sees a man or a longboat he goeth underwater
and runs away. I never did see nor hear, that out of his
own malice he endeavour'd to hurt any man, but when
he is in danger.*[1]

—Frederick Martens, German whaler, 1671

Mariners first sighted the towering, snow-capped peaks of
Spitsbergen on June 17, 1596. Like so many discoveries of the
period, this one was an accident. The pair of Dutch sailing ships
was actually heading for the unexplored coast of Siberia in
search of a Northeast Passage to the fabled Kingdom of Cathay.
Piloted by Willem Barents and financed by a group of Amsterdam merchants, the expedition wandered far to the north of the
usual sea lanes and had crossed the eightieth parallel of latitude,
farther north than anyone had been before them, when they
stumbled upon the high unknown coastline. "The land . . .
consisted only of mountains and pointed hills," wrote Barents,
"for which reason we gave it the name of Spitsbergen" (from the
Dutch word *spits*, meaning pointed and *bergen* meaning hills).[2]
But so inadequate was Europe's knowledge of the northern
ocean that for many years the new discovery was thought to be a
part of Greenland.

No one had much use for a cluster of frozen mountains so
far to the north. As a result, Spitsbergen was ignored until 1607
when another expedition, this one British, discovered it all over

29

again. Henry Hudson was returning from an unsuccessful attempt to sail his eighty-ton ship *Hopewell* to the North Pole when he spied the coast. Hudson failed in an attempt to penetrate the ice that crowded down onto the north side of the island, but when he reached England he was able to report that the waters along the west coast teemed with giant whales. The English did not know whaling at this time, but they knew that the Basques did and a group of London merchants made arrangements to recruit whalers in the Basque country to accompany British ships to the Spitsbergen grounds.

The whales that migrated past the shores of Spitsbergen were not the right whales so familiar to the Basques from their hunting expeditions in the Atlantic. They were a different species altogether. The whalers called them the Greenland whale; today they are known as the bowhead, *Balaena mysticetus*. The bowhead grows to a maximum length of about sixty feet and may weigh up to one hundred tons. A distinctive curving jawbone is shaped to accommodate the largest mouthful of baleen of any whale. An average specimen produced as many as seven hundred pieces of the material, some of which measured up to twelve feet in length. Later this substance was used as a kind of pre-industrial plastic to manufacture a bewildering array of products; however, when bowhead whaling began in the seventeenth century, baleen was in only limited demand, for ornamentation. Much more valuable was the layer of oil-rich blubber, extra thick to insulate against the loss of body heat in the frigid Arctic waters. In fact, so effective is the blubber that if the slow-moving whale travelled much faster it would not be able to lose the heat generated by the extra activity and it would expire of heat prostration in the middle of the ice. This coat of blubber produced a high-quality oil that was used in lamps as an illuminant and, just as important, as an ingredient in the manufacture of soap.

The Spitsbergen whales inhabited the open drift ice that collects along the edge of the solid ice pack surrounding the North Pole. In the winter they retreated south ahead of the expanding pack, only to return north to the coastal fjords of Spitsbergen with the spring breakup in search of food. When sunlight penetrates the surface of the water, large patches of plankton bloom in the ocean's upper layer at the ice edge. The plankton provides food for swarms of tiny crustaceans, or zooplankton. It is on these small animals that the whales feed, and so it was toward the ice edge that the whalers brought their ships.

I

The British were ready to begin whaling at Spitsbergen in the spring of 1611. The Muscovy Company, London's first great chartered trading company, dispatched two whaleships into the Arctic under the command of Jonas Poole.[3] Six Basque harpooners accompanied the expedition and managed to catch thirteen bowhead, the first whales killed commercially in the Arctic. But the venture ended in disaster. While Poole was away hunting walruses in one of the ships, the other was lost in the ice on the west coast of Spitsbergen. The crew was rescued by another ship from England, commanded by Thomas Marmaduke, who salvaged the wreck and some of the oil the men had been boiling on shore. When Poole arrived on the scene he reclaimed the blubber, but as he was loading oil into his vessel he carelessly mishandled the ballast and the ship went over on its side and sank. All the sailors reached safety, and Poole suffered the embarrassment of having to accept a ride back to England with

Boiling the blubber.

Marmaduke, who added to the humiliation by charging for the favour.

Despite the loss of two ships, the capture of so many whales encouraged the Muscovy Company and in 1612 another pair of vessels went out to Spitsbergen, with Jonas Poole in command once again. This time they killed seventeen whales and a number of walruses without mishap. "The whales," remarked Poole, "lay so thicke about the ship that some ran against our cables, some against the ship, and one against the rudder. One lay under our beake-head and slept there for a long while."[4]

However, the English were destined not to have this rich whaling ground to themselves. News of the venture had spread, and Poole encountered several other vessels at Spitsbergen that summer, including one from Holland and another from the Basque country. He shooed these interlopers away from the coast, claiming Muscovy Company ships had sole right to the whaling and, for now, the others obeyed. But this confrontation was merely the first skirmish in a long war for predominance in the Arctic.

The next season, 1613, the Muscovy Company armed itself with a charter from the English king, James I, providing a monopoly over the whale fishery and excluding all other vessels, English and foreign. A fleet of seven ships sailed to Spitsbergen, including the heavily-armed *Tiger*, sent to protect the whalers. Whales visited the coast in great numbers, as did ships from Europe determined to flout the English monopoly. The *Tiger* came upon the first Dutch vessels early in June and ordered them away from the island. The Dutch seemed to comply, but English commander Benjamin Joseph found them a few days later with their boats out. Angrily, he readied his guns and warned the interlopers that if they did not depart he would fire on them. This seemed to convince the Dutch, who handed over the blubber from eighteen whales and went on their way. Meanwhile, the ports of Biscay had sent fourteen whaleships north to Spitsbergen and Joseph had to intercept them all. Some he told to leave; others he allowed to continue whaling as long as they divided their catch with the English. The Basques were still the most proficient whalers in the world and doubtless the English hoped to learn from them.

The season of 1613 was a turning point in the Spitsbergen fishery. When the European whaling nations sent representatives

Arctic whaling scene. This early woodcut gives a hair-raising impression of the perils of northern whaling. The wild animals in the foreground are totally fictitious, though the whaling is shown accurately enough.

to England demanding restitution for the high-handed actions of Commander Joseph, they received no satisfaction. It was clear that if Holland and the others wanted a share of the action they were going to have to use force to secure it. As a result, a consortium of Dutch merchants, realizing that strength lay in unity, launched their own whaling company, the Noordsche Compagnie, and acquired a monopoly from the Netherlands' central parliament, the States-General. In the spring of 1614 the new company sent fourteen ships to the whaling, guarded by four men-of-war with thirty guns apiece. No longer were the Dutch whalemen willing to retreat passively in the face of English threats. Events were moving toward a showdown.

At first the Dutch show of force cowed the English whalers. It turned out to be a disastrous year for the hunt. The whales arrived on the coast late in the season, temperatures remained cold and the ice clung to the shoreline and blocked the fjords for the entire summer. Combined with the new Dutch attitude of defiance, the failure of the hunt robbed English whalers of some of their arrogance and on June 23 the two sides agreed to divide the west coast of Spitsbergen between them. The English received exclusive rights to four important harbours that experience had shown to be productive hunting grounds. The Dutch received free access to the rest of the coast. Both fleets declared that they would drive away whaling vessels belonging to any other nation.

Over the next three seasons this agreement dissolved in a series of minor squabbles. In 1617 the English, exasperated at the Dutch refusal to stop trespassing in English harbours, seized one of their ships and removed its cargo of oil and some of its whaling equipment before ordering it home. Once again the Dutch vowed to retaliate. This time they meant it.

The next season the Dutch arrived at Spitsbergen well armed and in a foul temper. As their men-of-war stood watch, they sent boats of armed men out on the water to protect the whalers as they went about their business. At Sir Thomas Smith's Bay, always recognized as a British harbour, the Dutch stationed a small flotilla of vessels, daring their rivals to start something. The British admiral, William Heley, had two whaleships and a small pinnace under his command. When the Dutch arrived in the bay, Heley sent a message asking to speak to their commander, who answered that he was otherwise occupied. Continuing with his bluff, Heley went ashore where the Dutch were setting up their equipment and haughtily ordered them to leave the harbour. The response was contemptuous. "You thinke to doe as you did the last yeare," sneered one of the Dutch crew, "but we are fitted for you nowe, and wilbe even with you for the last yeares work."[5]

Finally Heley came face to face with Captain Huybrecht Cornelisz, the Dutch Commander. It was an unsatisfactory meeting. The Dutchman was drunk and attacked his English counterpart with a drawn knife. Crewmen separated the two men before any harm was done, but the Dutch went away warning the British that before the season was over there would be more violence.

After this turbulent beginning the season passed calmly enough. But the Dutch were merely biding their time and waiting for more of their ships to arrive. When he felt strong enough, Cornelisz summoned Heley and the other British captains and ordered them to hand over their oil and equipment or he would take it by force and sink their vessels. Heley refused, and soon after the Dutch attacked with cannon blazing. An English eyewitness left an account of the skirmish. "We then, beeinge unable to make resistance againste so manie, they came aboorde of us, armed, and disarmed our ship of all her ordnance, powder, and munition, commandeinge our men to goe ashoare, pilladgeinge everye thinge they could laie hands on, drinkeinge out our beere, carreinge away our victualls, and doeinge what pleased themselves."[6] After loading the casks of oil and bundles of whalebone, and stealing whatever food, liquor and whaling supplies they could find, the Dutch sailed away, leaving their competitors to bind up their stricken ships and get home as best they could.

After the violent season of 1618 the whalers seemed to recognize that they could not whale and fight at the same time. The usual diplomatic wrangles produced an agreement that allowed Dutch and English ships to share the Spitsbergen grounds. Such agreements had been signed before, but this time the whalers themselves were tired of the squabbling, which if nothing else was cutting into their profits, and they arrived at their own *modus vivendi*. The Dutch installed themselves at the northwest corner of the archipelago and along the north coast, while the British remained at the familiar harbours of the west coast. This arrangement marked an end to the violent, opening

The whale hunt in Spitsbergen in 1630.

phase of Spitsbergen whaling. It survived for as long as whalers visited the coast. When other countries got involved in the Arctic fishery–the Danes, the French and the Germans–they found space for themselves at harbours which neither the Dutch nor the English used.

II

Whaling at Spitsbergen followed the model pioneered by Basque whalers in America. When whalers arrived on the coast in May they anchored their ships in protected coves near the best hunting grounds, then went ashore to set up their cauldrons on the beach in readiness for boiling the blubber. While some of the crew went away in the boats to patrol the coast in search of whales, the rest remained on shore tending the furnaces and transporting the barrels of oil out to the ships. By August the whales were leaving the coast, so the whalers packed up their equipment and departed for home.

Robert Fotherby, an officer with the English fleet in 1613, left the earliest account of whaling at Spitsbergen:[7]

When he entres into the sounds, our whal killers
doe presentlie sallie forth to meet him. Then,
comeing neare him, they rowe resolutlie towards
him, as though they intended to force the shallop
[the same boats that the Basques called *Chalupa*]
upon him. But so soone as they come within stroak
of him, the harponier, (who stands up readie in the
head of the boats,) darts his harping iron at him
out of both his hands, wherwith the whale being
stricken, he presentlie discends to the bottom of
the water, and therfor the men in the shallop doe
weire out 40, 50, or 60 fathoms of rope–yea,
sometimes 100, or more, according as the depth
requireth. For upon the sockett of the harping iron
ther is made fast a rope, which lies orderlie coiled
up in the sterne of the boat, which, I saie, they do
weire forth untill they perceave him to be riseing
againe, and then they haile in some of it. . . .

When they were able to get close enough to their prey, hunters
delivered the death blow with a long, sharp lance that they
plunged into the animal's body, seeking to pierce a major artery
or better yet, the heart or lungs.

And now will he frisk and strike with his tail very
forceablie, sometimes hitting the shallop, and
splitting hir asunder, sometimes, also, maihmeing
or killing some of the men. And for that cause,
ther is alwaies two or 3 shallops about the killing of
one whale, that one of them maie relieve and take
in the men out of another, being splitt.

Boats towed the dead whales back to their mother ship
where men with sharp blades cut the flesh and blubber into
large, rectangular pieces. Each piece was torn free of the carcass
using a hook attached by a rope to the capstan on the ship's
deck. Then the pieces were lowered into the water where, side by
side, they formed a raft which eventually was towed ashore for
boiling.

On the beach, choppers cut the blubber into small pieces
which were ladled into the huge cauldrons. Once the oil
appeared ready, it was scooped out of the cauldron, strained

through a wicker basket to remove pieces of skin and other impurities, and drained into an empty shallop. "And this shallop," Fotherby continued, "because it receaves the oile hott out of the two coppers, is kept continuallie half full of water, which is not onlie a meanes to coole the oile befor it runnes into the cask, but also to cleanse it from soot and drosse, which discends to the bottome of the boat. And out of this shallop the oile runneth into a long trough, or gutter of wood, and therby is conveyed into butts and hogsheads. . . ." While boiling was going on, other members of the crew removed the animal's head, brought it ashore and removed the pieces of baleen from the jaw with hatchets. Each piece was dried, scraped, rubbed with sand to clean off the grease, then bundled up for storage in the ship's hold.

After a few seasons it became obvious to the whalers that instead of dismantling their furnaces each season it would be more efficient to erect permanent boiling houses and leave equipment and boats for the winter. The Dutch located their blubber works on a beach at Amsterdam Island off the northwest corner of Spitsbergen, while the English based themselves at harbours to the south. The tiny settlements consisted of brick furnaces with attached chimneys, warehouses for storage, huts for the men and cooperages where the casks were made and repaired.

The Europeans hoped to colonize these Arctic whaling stations, at least with enough settlers to discourage pilfering by rival whalers. But it was not easy to find people desperate enough to move willingly to such a forbidding spot. The Muscovy Company recruited a group of criminals facing the death sentence and transported them to Spitsbergen with promises of food, shelter and a judicial reprieve if they stayed a year. The convicts took one look and declared to a man that they would rather hang than remain for the winter.

The Dutch had better success finding Arctic winterers, but the results were appalling. In 1633 the Noordsche Compagnie landed seven men at its whaling station on Jan Mayen Island in the Greenland Sea southwest of Spitsbergen. As the sun fell permanently below the horizon, the sojourners dined on salt meat and drank snow water melted at their fire. Polar bears wandered freely through the camp but the Dutch guns proved ineffective at bringing the huge animals down. The need for fresh food to ward off scurvy was not understood and by mid-March all seven men were so sick that they could barely walk. On April 16, Easter, the first man died. The rest had run out of food. "We were by this time reduced to a very deplorable state," wrote the man responsible for keeping the journal, "there being none of them all except myself, that were able to help themselves, much less one another."[8] On the last day of April the journal stopped. When a relief vessel arrived a short time later it found every member of the tiny colony dead.

That same year, at their other station located on Spitsbergen, the Dutch left another seven winterers. These men survived without mishap, feasting on reindeer, polar bear and fox meat. Unhappily, the venture was repeated the winter of 1634-35. Whether through their own ineptitude or some change in the climate, the men failed to supply themselves with country food. By the end of November, scurvy had already made its appearance. During January three men died. The journal entry for February 26 read: "Four of us that are still alive lie flat upon the Ground in our Hutts [that is, bunks]. We believe we could still feed, were there but one among us that could stir out of his Hutt to get us some Fewel, but no Body is able to stir for Pain. We spend our time in constant Prayers to implore God's Mercy to deliver us out of this misery, being ready whenever He pleases to call on us."[9] Then the pen fell from weakened fingers and silence descended on the camp. When the first ship arrived in the harbour that summer one of the sailors hurried ashore, broke open the door of the house and, running upstairs, stumbled over the frozen bodies of the last four winterers.

III

It was in the summer of 1617, on a flat stone beach in a bay at the southeast corner of Amsterdam Island, that Dutch whalers began building this station, planning for it to become their main shore station at Spitsbergen. Whalers were plentiful along this stretch of coastline and each season a few more structures appeared until the settlement seemed large enough to require a name; they called it Smeerenburg, "Melting City," site of the glowing furnaces that rendered the blubber into golden oil.

By the 1630s Smeerenburg had blossomed into the world's northernmost city. In the winter it lay deserted, its buildings buried beneath a mantle of snow, the only sound the thunder of ice floes clashing in the harbour and the whistle of the arctic

wind through the empty chimneys. But in the summer, with the
arrival of the whaling fleet, Smeerenburg came alive. All around
the shores of the bay, partners in the Noordsche Compagnie
built their warehouses, cookeries, workshops and dormitories.
At the waterfront, boats hurried back and forth between shore
and the long line of ships anchored in the harbour. Blacksmiths
bent to their forges making harpoons and honing the flensing
spades to razor sharpness. Coopers knocked the casks into
shape. Pedlars noisily hawked their wares among the sailors
loitering along the beach. Even a baker came to the island, and
the smells from his oven were a welcome relief from the stench
of burning whale skin.

A wooden fort bristling with cannon stood sentinel at one
end of town. For sailors with a few hours of leisure there was a
church and a tavern, but leisure was in short supply at Smeerenburg, a town that never slept. Night and day crews stoked the
blazing furnaces, boiling up the oil, filling the casks, then laying
them down in the ships' holds. The season was short; not a
minute could be wasted if the whalers hoped to make a profit.

Smeerenburg flourished during the 1630s and early 1640s
when its summer population may have reached several thousand
whalers and artisans. There was no comparable settlement in the
whole of Spitsbergen. But then, as quickly as it appeared, the
northern whale town faded away. The bowhead began to
disappear from the coast. Amsterdam Island lost its convenience
for Dutch whalers, as they extended their cruises farther from
shore in search of their prey. At Smeerenburg, the cauldrons
were carried away and the ovens pulled down. Buildings decayed
and collapsed. Only the occasional ship put in to make repairs.
In 1671, when Frederick Martens, a German whaler, visited the
site, the only reminder of Smeerenburg's glory days was a
scattering of dilapidated buildings, most of them gutted by fire.

IV

In 1642 the Noordsche Compagnie lost its monopoly and the
Spitsbergen grounds were thrown open to ships from any of the
Dutch ports. Holland at this time was embarking on a "golden
century" of commercial expansion. A long war with Spain was
drawing to a close, and the seven United Provinces of the Free
Netherlands, led by the coastal provinces of Holland and
Zeeland, were emerging as the dominant economic force in

Europe. Dutch expansion was based on the sea, on colonial trade with the Far East, on the carrying trade between European countries, and on the fisheries in the North Sea. As a result there were plenty of ships and trained sailors eager to take up Arctic whaling when the Noordsche Compagnie charter lapsed. Whereas the company in its heyday had never sent more than a few dozen vessels to Spitsbergen, by the 1670s about three hundred Dutch whaleships joined the hunt.

The decline of Smeerenburg marked a fundamental shift in the technique of Arctic whaling in the middle of the seventeenth century. As fewer whales visited the coast of Spitsbergen, Dutch whalers themselves moved away from the archipelago, seeking their prey near the ice edge on the high seas. Hunting so far from land, whalers no longer took the time to bring their blubber to shore for boiling. A Basque whaling captain, Martin Sopite, gains credit for being the first whaler to boil blubber on the deck of his ship over a small brick furnace called a tryworks.[10] But it was another century before American whalers utilized the full potential of these sea-going ovens. Early tryworks were small and presented a serious fire hazard on an oil-soaked ship. The Dutch preferred to remove the blubber, then pack it in barrels for boiling when they returned to port.

As the hunt moved out onto the high seas, Dutch whalers confirmed their leadership of the industry. A handful of ships from other countries continued to visit the Arctic each summer, but nothing like the huge Dutch fleet. The English failed to make the transition from shore to deep-sea whaling and were completely eclipsed when the bowhead began to disappear from the harbours along Spitsbergen's west coast. By 1669 English whale merchants sent only one vessel to sea, at the same time as the Dutch sponsored a mighty fleet of between two and three hundred ships.

When the seventeenth century began, the Dutch knew absolutely nothing about whaling. It was the Basques who were the acknowledged experts. Yet as the century drew to a close, the Dutch were the foremost whalers in the world. They had learned from the Basques, then driven them from the business. Then they had outfought their British rivals. Dutch oil and baleen flooded the markets of Europe. Dutch merchants owned 70 percent of all whaleships pursuing the hunt. It was an impressive achievement. But it was destined to be shortlived.

PLOUGHING THE ROUGHER OCEAN

If these people are not famous for tracing the fragrant furrow on the plain, they plough the rougher ocean, they gather from its surface, at an immense distance, and with Herculean labours, the riches it affords; they go to hunt and catch that huge fish which by its strength and velocity one would imagine ought to be beyond the reach of man.[1]

–Hector St John de Crèvecoeur, *Letters from
an American Farmer*, 1782

When the first American colonists arrived on the green shores of New England, they exclaimed at the hordes of whales cavorting in the coastal waters. "It was a place of profitable fishing," was how the Pilgrims aboard the *Mayflower* described their new homeland in 1620, "for large whales of the best kind for oil and bone came daily alongside and played about the ship."[2] Lacking the know-how to pursue the animals at sea, early colonists had to be content to salvage oil from dead whales stranded on the beach. By the 1640s they were venturing out in open boats to hunt right whales migrating along the coast. These amateurish crews of landsmen represent the beginning of the American whaling adventure. Over the next two centuries their small boats grew into a huge armada of sailing ships, spreading their white canvas across the oceans of the globe, seeking whales wherever they congregated and opening undiscovered worlds to the mixed blessings of Western civilization.

I

Now a fashionable suburb of New York City, Long Island can fairly claim to be the cradle of the American whaling industry. In the 1640s, shortly after the first settlers came to clear the forests at the eastern end of the island at Southampton and East Hampton, residents began organizing themselves into patrols during the winter months when whales were known to pass alongshore. Spotters stationed on bluffs overlooking the beach raised the alarm when they spotted an animal blowing out beyond the breaking surf. Hurrying from their houses, the men of the community ran down to the water's edge and launched their boats through the waves. With the lookout signalling with a flag attached to a long pole, indicating the location of the whale, the six-man crews pulled their oars in hot pursuit. After the kill, the boats towed the animal back to shore where the townsfolk butchered it and boiled the blubber in iron kettles on the beach.[3]

Every able-bodied man was expected to join these expeditions, and local Indians were hired as well. Each household received a share of the catch, proportional to what it contributed to the hunt. During the whaling season, roughly November to

Whaling off Long Island. Shore whalers remove the blubber from their catch. From Harper's Weekly, *Jan. 31, 1885.*

April, even the schools closed so that children could help with the trying out. Teachers didn't care; they received their salary in oil and baleen. Along with farming, whaling grew to be the economic mainstay of the communities on Long Island, and gradually it spread up and down the Atlantic seaboard.

Toward the end of the seventeenth century the residents of Nantucket took up whaling. Named after an Indian word meaning "land far out to sea," Nantucket lies a half-day's sail off the south coast of Cape Cod. First settled in 1659, it is a low, bleak, sandspit of an island, fourteen miles from end to end, dappled with ponds and swamps and covered with stiff grasses and stunted vegetation. The original settlers, who purchased the island for thirty pounds sterling and a pair of beaver hats, were religious nonconformists seeking the freedom that isolation brings. Technically, their island belonged to New York, but it was years before anyone in authority paid any attention to them, and then it was only to collect an annual quitrent of four barrels of fish. Islanders called the mainland simply "America," and did not feel themselves a part of it.

Unable to support themselves by farming the sandy soil, they brought sheep with them, hoping to export the wool to the mainland in exchange for the provisions they needed to feed their colony. The inadequacy of an economy based on herds of scrawny sheep was dawning on Nantucketers in 1672 when a large whale strayed into the harbour at Sherburne, the island's main settlement. Quickly forging themselves a harpoon, residents attacked and killed the animal. From this modest beginning grew an industry that carried the name of Nantucket to the farthest reaches of the globe.

It was no accident that Nantucket islanders took to whaling with such enthusiasm. Nantucket had no natural resources of any importance. The people quite naturally looked seaward for their livelihood. Their offshore location placed them close to the migrating whales, and an excellent harbour gave shelter to their growing fleet of vessels. Experts came from the mainland to teach the islanders how to come down on a whale without frightening it, where to thrust the deadly harpoon, and how to process the blubber to obtain the finest-quality oil. Before many years had passed, life on the island was revolving around the fishery.

Hector St John de Crèvecoeur, in his classic *Letters from an American Farmer*, described how islanders methodically trained their children in all aspects of the business. First the youngsters went to school to learn to read and write and do their calculations. Then, at age twelve, they apprenticed to the barrel-making trade. At fourteen they were ready to go to sea. "They learn the great and useful art of working a ship in all the different situations which the sea and wind so often require," wrote de Crèvecoeur, "and surely there can not be a better or more useful school of that kind in the world. Then they go gradually through every station of rowers, steersmen, and harpooners; thus they learn to attack, to pursue, to overtake, to cut, to dress their huge game: and after having performed several such voyages, and perfected themselves in this business, they are fit either for the counting-house or chase."[4]

In 1702, a Quaker missionary paid a visit to Nantucket. At a gathering of the curious in the home of one of the leading families, the Society of Friends made its first island convert. From this emotional beginning, the austere Quaker faith took firm root in Nantucket's sandy soil. It increased the homogeneity of island society and the sense of uniqueness that separated the people from the mainland. In times of war, when their pacifism earned the suspicion of other Americans, their isolation was even more acute. At the same time, Quakers preached the virtues of industry and frugality so necessary to a community dependent on trade and commerce. Visitors to the island commented on the bustle of activity in the narrow, cobble-stoned streets, the simplicity of the grey-shingled houses, the plain dress of the citizens, the sobriety of their manner. It was a notable marriage of religion and commerce, and for a hundred years it made the tiny island one of the most prosperous communities in the world.

Nantucket moved to the forefront of the young American whaling industry with the transition from along-shore to deep-water whaling. As the number of whales close to land declined, hunters ventured farther from the coast and stayed away two or three weeks at a time. As the voyages got longer, the vessels grew bigger. Whalers began to use fifty- to ninety-ton sloops with crews of thirteen sailors to carry a pair of boats out into the mid-Atlantic.[5] Instead of towing dead animals to shore, crewmen stripped the carcasses on the spot, put the blubber into barrels and carried it back to port for rendering.

About 1712 a Nantucket captain named Hussey made a historic voyage. According to local historian Obed Macey,

Hussey was hunting right whales close to shore when a gale blew his sloop out to sea. As the weather cleared, Hussey spied a whale swimming nearby and attacked it, perhaps without knowing exactly what it was. The animal turned out to be a sperm whale, a species not often seen near land and much feared by whalers for its size and fierce temper. Hussey managed to kill the huge beast, showing that it could be done and inspiring a growing number of whaling captains to follow him out into the deep.

II

Once they became familiar with the sperm's habits, no whaler could confuse the right whale with its deep-water relation. The sperm whale feeds on giant squid far below the ocean surface and can remain underwater for more than an hour. Its blubber produces less oil than the right whale's, but the best sperm oil was worth three times as much because it was lighter and burned with a pure, odourless flame. Unlike the right whale, the sperm is a toothed whale and carries no baleen in its mouth. Its huge, blunt head, taking up as much as a third of the animal's total length, contains a unique receptacle called the case. Inside the case is a mixture of very high quality oil and a wax-like, fatty substance known as spermaceti. Spermaceti can be separated from the rest of the head matter, and by the 1750s it was being used to make a new kind of candle that produced a brilliant, almost smokeless flame.

Historians do not agree about the inventor of the spermaceti candle. Some give credit to Benjamin Crabb, a chandler who opened a shop at Providence, Rhode Island, in 1751. Others claim it was the Sephardic Jew, Jacob Rodriguez Rivera, who brought a secret process to Newport, Rhode Island, whence it spread to other New England chandleries. Regardless, the candles became so popular with householders that within a few years Americans were exporting hundreds of thousands of pounds' worth to Europe and the West Indies.[6]

Though shore whaling continued close to the New England coast, Nantucketers began to concentrate on the sperm whale hunt, venturing farther out into the Atlantic in their small sloops and schooners. Cruising north, Yankee whalers reached the

Grand Banks off Newfoundland, entered the Gulf of St Lawrence and penetrated up into Davis Strait. To the south, they visited Bermuda and the Bahamas Banks. As their ships increased in size, whalers no longer needed to return to shore with every whale they killed. But they still had to bring home their blubber periodically for boiling before it spoiled. In the 1750s this final link to the land was broken with the introduction of tryworks on board the vessels. First invented by the Basques a hundred years earlier, a tryworks was a brick furnace mounted on the deck of the whaling ship. A fire in the works heated a large iron cauldron full of blubber. As the blubber boiled, oil separated from the fibres. When it reached the proper quality the oil was scooped into barrels and stored in the hold for the rest of the voyage. With this innovation, whalers were free to sail as far as they wanted in search of their prey.

Sperm whalers used pretty much the same techniques that Basque whalers perfected in the Bay of Biscay and the other European whalers employed in the ice near Spitsbergen. Ships grew larger, voyages got longer, but men still carried on the hunt with hand-held weapons in small boats about half the length of their prey. The contest was fairly equal, and was just as likely to end in a smashed boat as a dead whale.

The hunt began with the familiar cry from the masthead: "There she blows!" Crewmen dropped whatever they were doing and raced to the rails to scan the blank horizon. "Where away?" the captain wanted to know as he emerged on deck. "How far off?" On a clear day in a calm sea a sharp-eyed lookout could spot a whale spouting several miles away and usually could identify the species from the shape and frequency of the spray. A sperm whale's blow is low and "bushy" and angles forward. It generally spouts thirty to forty times when it surfaces and the intervals between blows do not vary, allowing the spotter to "take the tenor of the spout" and confirm the identity of the whale.

When the ship closed to a reasonable distance, the boats dropped into the water to take up the chase. The crew of each boat prepared their craft with care for this moment. Harpoons, lances, knives and axes were sharpened and stowed in their proper places. The long whale line lay smoothly coiled in a tub on the floor of the boat. Bailing buckets, emergency rations, a knock-down sail, extra oars, a keg of water, a lantern were all on board. The earliest deep-sea sloops only brought along a pair of

boats. These were square-sterned craft, crewed by three oarsmen, a harpooner and a boatheader working the tiller. As the
business evolved, so did the boat, emerging by the time of the
American Revolution as a sleek, double-ended craft up to thirty
feet long, with a crew of five oarsmen plus the boatheader.[7]

Once the boats were away across the water the whale had
usually sounded. It was the header's job to steer to the spot
where the animal was likely to come up for its next breath.
Wordlessly, sometimes for hours without a break, the men bent
to their oars. They faced away from the whale and only knew that
they had caught up with it when the header sang out "Stand up!"
and the harpooner put aside his oar and braced himself in the
bow where the harpoon lay ready to hand.

The harpooner's job required incredible strength. Not only
did he have to handle an oar like any other member of the crew,
he then had to stand up in a heaving boat and propel a heavy
harpoon with enough force to plant it deep in the whale's back.
Originally, whalers used a two-flued, or twin-barbed, harpoon.
As experience showed that these often worked their way out of
the fleeing animal, the single-flued "iron," which bent and
twisted without pulling, gained popularity. Finally, in 1848,
Lewis Temple, a former slave who settled in New Bedford as a
blacksmith, developed the toggle harpoon with a head that
swivelled on its shank. This iron almost never withdrew and one
writer has called it "the most important single invention in the
whole history of whaling."[8]

If the harpooner was successful with his first thrust, he
quickly tried to fasten a second iron in the huge beast, while the
header bellowed "Stern all!" and the oarsmen frantically backed
the boat away to avoid a convulsive swipe of the whale's tail
flukes. Then, enraged with pain and fear, the animal surged
away—"the poles sticking up in his back like stray hairs on a bald
head," wrote one whaler—dragging line, boat and crew behind it
in a mad dash for freedom. It was a dangerous moment. A
careless crewman getting in the way of the unwinding rope was
instantly entangled and yanked overboard. If he could not cut
himself free, the unfortunate seaman was dragged beneath the
water and drowned, his waterlogged body only returning to the
surface if the animal was killed. Others were badly mutilated by
the rope, which, moving so quickly without any slack, sliced
through tissue and bone like a sharp knife. The wooden leg and
the missing hand were not just the melodramatic props of

fiction; every whaleship had crewmen who had lost some part of their bodies to the whale's fury.

The thick whale line whirred out of the tub near the rear of the boat. It looped around the wooden loggerhead beside the tiller, then forward along the line of the keel between the oarsmen and over the bow into the black depths. The slightest kink in the line caught on the boat and dragged it under. The line wrapped around the loggerhead to provide enough tension to tire the whale, much as a fishing line wraps around a reel. It was the job of one of the oarsmen to keep the rope wet so that it would not burst into flame from the friction.

The whalers played their catch much like a fish on a hook. The crew sat facing forward, hauling on the line as the boat flew over the water in the celebrated "Nantucket sleighride," reaching speeds up to twenty-five miles an hour. When the whale began its climb back to the surface, the men felt the rope slacken and hauled in hand over hand, trying to shorten the distance between themselves and the wounded animal. Meanwhile, any other boats in the vicinity rushed to the spot to assist the "fast boat" in the kill. It was unusual for a whale to take out the entire line, but some did and line from a free boat had to be tied on to continue the chase.

As long as the whale stayed strong and angry the boats stood off at a respectful distance. Eventually the exhausted animal weakened and a boat worked in close enough to administer the fatal blow. By this time the harpooner had changed places with the header, who reserved the honour of the kill for himself. Taking up a long, spear-like lance, he plunged it again and again into the whale's vitals until it pierced the heart or severed a large vein. In whaleman's language, this was known as "tapping the claret bottle." The blowhole spouted crimson blood (called "running up the red flag" or "setting his chimney ablaze") and the whale went into an agonizing death flurry, swimming in tight circles, thrashing the water into a foamy tempest and vomiting half-digested chunks of food. At last, its lifeblood staining the water all round, the great beast rolled over and lay still.

Once they retrieved their lines and caught their breath, the whalers had to get the huge carcass back to the ship. If they were lucky, the vessel could sail down and take the whale alongside. But in calm weather, or when the chase carried the boat some distance away, sailors attached a rope and took the carcass under

Flensing. A long strip of oil-rich blubber is hauled on board a Pacific whaleship. Smoke from the tryworks is blackening the sails.

tow. This return trip, with forty tons of deadweight behind, could take several hours.

Flensing began as soon as the animal was lashed alongside the ship. One of the boat-steerers clambered onto the slippery back to cut a hole in the skin and blubber near the whale's eye. A hook, inserted into the hole, connected through a series of ropes and pulleys to the ship's windlass. As crewmen on board turned the windlass, and others on the whale cut with their sharp flensing spades, a strip of blubber one to two feet wide began peeling off the carcass. Dripping oil and blood, this "blanket piece" rose into the air until the two blocks of the tackle came together and no rope remained to lift the blubber higher. Then it was sliced free and lowered through a hatch into the blubber room below deck, while a hook was attached to the next piece. Slab by giant slab, flensers stripped the blubber from the carcass until only a bloody mass of bones and entrails remained.

Meanwhile, other crewmen severed the sperm whale's head and hoisted it on board. After removing the teeth from their sockets in the lower jaw, sawyers cut the jaw-bone into pieces. Both teeth and bone provided raw material for scrimshawing, the whalers' art of engraving elaborate scenes and designs.

Trying out. Blubber was rendered into oil by boiling in the tryworks. Here, the man in the centre is skimming debris off the surface of the oil.

The spermaceti was removed from the head case and stored separately.

Occasionally a lucky whaleship found a lump of ambergris, a waxy substance secreted in the lower intestine of the sperm whale. Weighing anywhere from a few ounces to several hundred pounds, this strange, amber-grey substance was valued as an aphrodisiac and as a fixative in the preparation of perfume. It turned up rarely—sometimes a large mass was found floating at sea—and it commanded a very high price. The largest recorded piece, found by a British whaler near New Zealand, weighed 1,400 pounds and was worth £25,000.[9]

Weather permitting, "trying out" began as soon as the "cutting in" was complete, before the blubber had a chance to decay. Crewmen armed with sharp cleavers cut the blanket pieces into small, rectangular "horse pieces" which in turn were sliced thinly, like the pages of a book. These pieces, called "bible leaves," were fed into the trypots and boiled until all the oil was extracted. Tenders skimmed off the residue of skin and tissue and used it to fuel the fire in the tryworks.

The trying-out process held a particular fascination for whalemen, some of whom described it like a scene from hell. "There is a murderous appearance about the blood stained decks," wrote J. Ross Browne, "and the huge masses of flesh and blubber lying here and there, and a ferocity in the looks of the men, heightened by the red, fierce glare of the fires, which inspire in the mind of the novice feelings of mingled disgust and awe."[10] Herman Melville drew on Browne's book *Etchings of a Whaling Cruise*, when he was composing his own whaling epic and the famous tryworks chapter in *Moby Dick* owes part of its inspiration to this riveting description by Browne:

We will now imagine the works in full operation at night. Dense clouds of lurid smoke are curling up to the tops, shrouding the rigging from the view. The oil is hissing in the try-pots. Half a dozen of the crew are sitting on the windlass, their rough, weather-beaten faces shining in the red glare of the fires, all clothed in greasy duck, and forming about as savage looking a group as ever was sketched by the pencil of Salvator Rosa. The cooper and one of the mates are raking up the fires with long bars of wood or iron. The decks, bulwarks, railings,

try-works, and windlass are covered with oil and
slime of black-skin, glistening with the red glare
from the try-works. Slowly and doggedly the vessel
is pitching her way through the rough seas, looking
as if enveloped in flames. . . .[11]

Trying out was the final step in the processing of a whale
carcass. Once the oil was securely stowed below, the fires were
extinguished, the decks and rigging scrubbed clean of oil, blood
and soot, and the crew fell back into the ordinary routine of
looking out for another whale.

III

At the same time as New Englanders moved cautiously out into
the Atlantic to chase the sperm whale, Arctic whalers, recognizing that bowhead whales were disappearing from the familiar
waters around Spitsbergen, began cruising in the open ocean
along the edge of the ice pack. At Spitsbergen the English and
Basques had rivalled the Dutch whalers, but once the hunt
shifted away from shore it became at once riskier and more
costly and the other nations reduced their efforts. By the end of
the seventeenth century the Dutch dominated the industry in
Europe. Each spring hundreds of their vessels went "ice fishing"
in the waters between Spitsbergen and Greenland along the
frozen hem of the polar pack, and Dutch merchants supplied all
of Europe with whale oil and baleen. More than 27,000 bowhead
were killed by Dutch whalers in the last four decades of the
century alone.[12]

Inexorably, the search for whales carried the hunters to the
ice-choked inlets of East Greenland, down the ragged coastline
to Cape Farewell, and around into Davis Strait, where large
numbers of bowhead were discovered. Now part of the Canadian
Arctic, this region to the west of Greenland was vaguely familiar
to European sailors. In the 1570s Martin Frobisher became the
first mariner to venture north of Labrador in search of a
Northwest Passage across the top of America to the Orient. He
thought he found such a passage on the coast of Baffin Island
where he entered a narrow inlet that seemed to lead away to the
west. Frobisher was distracted by a futile search for gold and
never did learn the truth of his discovery, that his "streyte" was
actually a deep bay, now Frobisher Bay. A few years later John

Davis arrived off the Baffin coast, also looking for the fabled westward passage. He had high hopes for Cumberland Sound, but it turned out to be another dead end. Then, in 1616, William Baffin sailed his tiny vessel *Discovery* through Davis Strait as far as the mouth of Sir Thomas Smith's Sound, farther north than anyone had been before him, and farther north than any other mariner would reach for another two hundred years.

Thanks to these early navigators, whaling captains possessed a general idea of the geography of Davis Strait when they began arriving in large numbers after 1720. The Dutch continued to predominate, but the discovery of a rich new whaling ground in Davis Strait rekindled British interest in the Arctic fishery. The demand for whale oil was strong, fuelled by a growing woollen textile industry that used huge quantities of oil as a wool cleanser. Oil was also in demand as a machinery lubricant, and for the new street-lighting systems that were spreading across Britain. The government decided to breathe some new life into the stagnant British whaling industry by subsidizing a growth in the size of the fleet. In 1733 the government announced that it would pay a bounty of twenty shillings per ton to every whaleship weighing more than two hundred tons. The bounty was soon raised to thirty shillings and then, in 1749, to forty shillings.[13] With this last increase the fleet was finally afloat, and the rebirth of British whaling is usually dated at the 1750s. By the end of the century, British mariners, supported by government subsidy, had completely eclipsed their Dutch rivals.

IV

British ships set sail for the whaling in Davis Strait by the end of April. As the quay, crowded with friends and family, fell away astern, the captain sent his mates to search for the inevitable stowaways who would have to be put ashore before the ship passed the last point of land. Under way at last, the canvas filling with wind, the captain made do with a skeleton crew for the first few days while his green hands lay moaning in their berths, limp with seasickness. It usually took three days for the full crew, between forty and fifty men, to get its sea legs.

On their way out, British whalers paused at the Shetland and Orkney Islands to take on extra hands. Incessant warfare between Britain and France so depleted the ranks of British seamen that the whalers had to seek their crews in out-of-the-way places like these bleak, treeless islands north of Scotland. But Shetlanders and Orkneymen proved so willing to take lower wages, and so adept at handling a whaleboat, that captains came to favour the islanders over other seamen.

While officers traded for fresh fish, chickens and produce from the locals, crewmen hurried ashore to the taverns of Stromness and Lerwick to take advantage of the last land they would touch before the ocean crossing. The island whisky shops were low, windowless huts made of mud and thick with smoke from the peat fires. "This shop was literally stowed with both sexes," reported one greenhorn after a visit to his first Lerwick tavern, "the greater part of them sailors, some singing, others swearing coarse oaths. In the centre of the den two of the tars were reeling with a dark-eyed island girl to the drone of a bagpipe, driven by an old and lank-jawed piper attired in a grey serge jacket and leaning against the smutty wall. Here was such a commingling of the low and the ridiculous as I had never before witnessed, a strange-looking lot, dimly visible through the veil of peat smoke, singing, roaring, yelling, dancing and snapping their fingers and stamping their feet."[14] Here the captains came to retrieve the "southern boys" and herd them back to the ship.

Whaleships laboured across the North Atlantic along the fifty-eighth parallel of latitude, bringing them to Greenland well south of Cape Farewell, a landmark much feared for the drifting ice that surrounded it. The crossing took ten days or ten weeks, depending on the wind and weather. Spring was not well advanced. Fierce storms swept the sea lanes and temperatures dipped well below freezing. Ocean spray froze on the rigging and coated the decks with a skin of ice. Large icebergs drifted past, visible even at night by their ghostly glow. More dangerous were the huge "washing pieces," chunks of ice riding so low in the water that they were indistinguishable in the breaking seas.

Down in the cramped quarters of the forecastle, the frozen sailors thawed themselves at the stove and dined on salt cod, potatoes, cheese, bread, and a cup of warming tea. Then the exhausted men climbed into their berths to snatch a few hours' rest before their next watch called them back on deck. When the weather cleared, the sailors amused themselves by catching fulmars. These grey-white birds—the whalemen called them "mollies"—joined the ships soon after they left the northern islands and followed them across to the whaling grounds.

Mountains of ice. Icebergs were a constant danger for whaleships venturing into the North. In this scene, sailors are forced to abandon their vessels, which are about to break apart under the pressure of the ice.

Smaller than a duck, mollies were scavengers, diving low to pluck from the water any bits of garbage thrown overboard. The men baited hooks with fat and trailed them in the wake on a piece of string to catch the birds, which made a nice change in diet from salt meat and fish. On calm, black nights the fulmars flitted across the decks on noiseless wings like bats, startling the sailors as they stood their watch.

Arctic whaleships challenged the northern ice with dimensions that are laughable stacked up against the giant vessels that venture north today. About one hundred feet in length, a whaler would fit comfortably on many suburban house lots. By comparison, the massive oil tanker *Manhattan*, which powered through the Canadian Arctic in 1969, was as long as three football fields and weighed 155,000 tons. A wooden whaleship weighed anywhere from two hundred to four hundred tons. Its hull was stoutly planked with two and sometimes three layers of boards, and bow and stern were fortified with thick oak beams and sheathed on the outside with wood or iron plates.[15] All of this was intended to withstand the relentless pressure of the pack as it closed around the vessel like a fist, and to cushion the inevitable collisions with the rock-hard ice. Still, for all these precautions, whaleships were defenseless against the awesome power of the ice, and it was a rare season that failed to see at least one ship crushed and sent to the bottom.

When the ships reached Cape Farewell crews began "spanning on." They removed the whaleboats from storage and hung them out on davits along the side of the ship. They honed the harpoons, knives, lances and flensing spades to razor sharpness and stowed them in their proper places, ready for use. Each boat carried a mile of heavy whale line that had to be attached to the harpoons, coiled carefully so that it would not tangle, then laid in the tub on the floor of the boat. When all the equipment was ready the crew assembled on the afterdeck to toast the success of the hunt with a glass of grog. The whaling had officially begun.

When whaleships rounded Cape Farewell into the mouth of Davis Strait they ran headlong into a stream of ice draining out of the Arctic. Each spring the ice that covers the inlets and sounds of the northern archipelago breaks up into countless floating pieces that drift southward on the current into Baffin Bay and on down Davis Strait. Because of the direction in which the currents flow, the ice is heaviest on the Baffin Island side of the strait early in the season and for many years no whaleship dared to cross to the "West Land." The east side of the strait, however, is washed by a warm current flowing up the coast of Greenland from the south. This current loosened the grip of the ice along the coast and early in the summer a corridor of open water stretched northward next to the towering Greenland shore all the way to Disco Island and beyond. It was up this passage that the whalers followed their prey.

Once on the whaling grounds, a lookout perched aloft in the crow's nest, scanning the horizon for spouting whales. Bowhead congregate near the edge of the ice, so the lookout watched for what the whalers called a "blink," a white reflection low in the sky indicating that ice lay in that direction. An experienced navigator could "read" the blink and tell by its shading whether the ice was field or pack, compact or open. Similarly, when a ship was hemmed in by ice floes, a bluish blink, called a "water sky," indicated open water in the distance.

The hunt for bowhead did not differ substantially from the sperm whale hunt, though the chase was complicated by the ice. A harpooned whale invariably sought to make its escape under the floes. If the ice stretched away in an unbroken field, whalers might let their quarry run with the line, content to wait at the ice edge where, unless it found a hole in the pack, the animal must return for air. If, on the other hand, the ice was drifting loosely in pieces, then the mate might allow his boat to be towed in among them. This was a very troublesome sea in which to chase whales. The boat was in constant danger of being smashed against the ice and the animals could not always be spotted when they surfaced. When loose ice packed together so that a boat could not follow the whale, the oarsmen leapt out of the boat with their harpoons and lances and, dancing from floe to floe, continued the hunt on foot. In time the animal would surface in one of the open spaces to catch its breath and when it did the whalers planted their extra irons. When a kill was made in the ice, the only way to retrieve the corpse was to sink it, either by cutting off its flukes or weighting it with sandbags, then to haul on the harpoon line until the carcass reached open water.

All of this took time, during which the dead animal was decomposing internally. Gases formed from decomposition were trapped by the blubber and the carcass could swell up like a huge balloon until it exploded in a shower of blood and tissue. Robert Goodsir, surgeon aboard a Davis Strait whaler, recalled being in a boat on his way to visit a nearby ship when he crossed paths with another boat towing a dead whale found drifting nearby. Distended with foul-smelling gases, the corpse emitted a peculiar sound as it moved through the water. Goodsir at first thought someone was playing the bagpipes, but as the decaying beast passed by he realized that each time the towboat jerked ahead, gas from the bowels of the carcass whistled out of the wounds in its flesh and "transformed the dead whale into a strange musical instrument."[16]

If seas were calm, the carcass was stripped of its blubber as soon as it was brought to the ship. The longer the delay, the greater the chance that the oil would be tainted by decay. Once the head, with its valuable load of baleen, was detached from the body and hoisted aboard, the blubber was removed in long strips, cut up into smaller pieces and stored below deck. Using wedges, crewmen separated the baleen from the jawbone of the giant head. Each piece would eventually be scrubbed clean, dried, bundled and stowed away. After flensing, the headless carcass, called the "kreng," drifted free of the ship to provide food for polar bears, seabirds and foxes. "Here and there along the floe edge lay the dead bodies of hundreds of flenched whales," recalled one whaler, "and the air for miles around was tainted with the faetor which arose from such masses of putridity."[17]

Flensing was a filthy business. The sides of the ship were splattered with blood and fat, the rigging slick with oil and the decks so grimy and slippery that it was almost impossible for sailors to stay on their feet. They sprinkled sawdust around the work areas to improve the footing, and immediately it was tracked into the cabins and forecastle so that the ship reeked with the stink of whale from bow to stern.

The cacophony of noise that accompanied the flensing of a whale is hard to imagine. A thick cloud of fulmars descended on the ship to gorge themselves on the blubber and oily waste. Flensers were constantly interrupting their work to brush away the persistent birds, sometimes slashing at them with their sharp cutting spades. "The noise they make at such times is almost deafening," reported Robert Goodsir, "and exactly resembles that of poultry, something between the cackle of the hen and the quack of a duck. . . ."[18] After filling up on fat, the birds drifted lazily on the current, too stuffed to fly or swim. Often they vomited the contents of their stomachs so that they could continue feeding. When flensing was over, an Arctic whaler was ringed with bloated fulmars contentedly sleeping off the effects of their orgy.

Arctic whaleships did not boil their blubber on board. Bowhead whales gave so much oil that the available tryworks were too small to handle it. Instead, blubber was chopped into small pieces that were put into barrels and carried back to European ports for trying out. On a southern voyage, which lasted a year or more, the blubber would have spoiled, but in the Arctic temperatures were cooler and ships were only away for about eight months. When September arrived the ice began to re-form and captains sped away home to avoid being trapped on the Greenland coast for the winter.

V

During the eighteenth century the whaling fleets of America and Great Britain were steadily extending their reach and increasing their size. The two fleets kept to different ends of the Atlantic—the British chased bowhead whales in the Arctic ice while the Americans had the southern sperm whale fishery all to themselves—but it was perhaps inevitable that they would begin to encroach on each other's preserves. The Americans turned out to be the more ambitious, visiting Davis Strait for the first time in 1732 and also whaling in the Gulf of St Lawrence and off Newfoundland. In 1765 British authorities began to think that the upstart colonial industry had better be restrained before it drove the mother country's fleet out of business. Claiming that Americans were committing "all Sorts of Outrages" on the coast of Labrador and Newfoundland, including despoiling the British fishing stations and mistreating the local Indians, naval authorities issued a set of restrictions that greatly inconvenienced the Yankee whalers. For example, they were ordered not to abandon their flensed carcasses within nine miles of shore; not to sign on any crewmen from among the fishermen on the coast; not to give liquor to the Indians; and not to fish for cod, even for their own consumption.[19]

While this policy succeeded in driving some of the New England whalers from the Grand Banks, it did not succeed in frustrating the overall expansion of the American industry. The fleet simply turned its attention to the south, spreading across the Atlantic to the Azores and then on to the coast of Africa where the Gulf of Guinea and Angola's Woolwich Bay became important grounds. In the 1770s American whalers opened new hunting grounds around the Falkland Islands and off the coasts of Brazil and Argentina. By the outbreak of the American Revolution, in 1775, the New England fleet had grown to 360 vessels; Nantucket alone sent out 150 sail.[20] While European whalers continued to make their predictable excursions into the northern ice, Americans were blanketing the Atlantic with their boats, from the glacier-clad shores of Greenland to the tip of Tierra del Fuego to the torrid sweep of sub-Saharan Africa.

Even the British were impressed at the energy of the Americans. During the parliamentary debates leading up to the outbreak of the Revolution, Edmund Burke delivered an eloquent tribute to the colonial whalers. "Whilst we follow them among the tumbling mountains of ice," he told his colleagues in the House of Commons, "and behold them penetrating into the deepest recesses of Hudson's Bay and Davis' Strait, whilst we are looking for them beneath the Arctic Circle, we hear that they have pierced into the opposite region of polar cold, that they are at the antipodes, and engaged under the frozen serpent of the south. We know," Burke continued, "that whilst some of them draw the line and strike the harpoon on the coast of Africa, others run the longitude, and pursue their gigantic game along the coast of Brazil. No sea but what is vexed by their fisheries. No climate that is not a witness to their toils. Neither the perseverance of Holland, nor the activity of France, nor the dextrous and firm sagacity of English enterprise ever carried this most perilous mode of hardy industry to the extent to which it has been pushed by this recent people; a people who are still, as it were, but in the gristle, and not yet hardened into the bone, of manhood."[21]

If Edmund Burke could afford to be admiring of the efforts of American whalers, British shipowners could not. Their country was already importing almost four times as much whale oil from its New England colonies as it received from its own Greenland fleet. Having just managed to surpass the Dutch, British arctic whalers faced a much tougher challenge from the Americans. Their industry appeared to be on the verge of collapse.

What saved the British feet was the outbreak of the American Revolution. As the colonies moved relentlessly toward a break with Great Britain, whalers in New England recognized that their business would be devastated by the conflict. Not only would their trade with England be cut off, but their ships would be easy prey for British naval cruisers and marauding privateers. On Nantucket, Quaker merchants were as concerned for their pacifist principles as they were for their economic self-interest. But the voice of the whalers had little influence with the leaders of the independence movement, and war came.

As expected, colonial whalers were beset on both sides. On the one hand British warships attacked unarmed whaleships and either impressed their crews into the British navy or forced them to go to work on ships in their own Arctic fleet. Closer to home, a British force landed at the Massachusetts whaling port of New Bedford and set the wharves and warehouses ablaze. Nantucket escaped a similar fate only because foul weather drove the attackers away. Meanwhile, the Continental Congress sitting in Philadelphia did not want the whalers trading with the enemy. Congress banned the export of whale products and required that shipowners post a bond before sending any vessel on a whaling voyage, a bond that would be forfeit if the ship did not return with its cargo to an American port.

Harassed at sea, mistrusted at home, forced to donate their ships to the war effort, American whalers abandoned the hunt. For three years whaling in New England ceased altogether. British whalers hurried to take advantage of the misfortunes of their rivals. Terminally ill just a few years before, the British industry rose from its deathbed and flourished for another fifty years. Instead it was the American fleet that faced an early death. "The bells that called the hardy yeomanry of New England to the defense of their imperiled liberties . . . rung the death-knell of the whale fishery," wrote one of the earliest historians of whaling, Alexander Starbuck. "The rattle of musketry was the funeral volley over its grave."[22]

WHALING
IN PARADISE

I freely assert, that the cosmopolite philosopher cannot, for his life, point out one single peaceful influence, which within the last sixty years has operated more potentially upon the whole broad world, taken in one aggregate, than the high and mighty business of whaling.

–Herman Melville, *Moby Dick*, 1851

The American Revolution turned out to be a setback for the Yankee whaling industry, but only a setback. In the years following the Revolution, American whalers confounded the doomsayers by bouncing back from near destruction to enjoy a period of unprecedented expansion. This was the era chronicled by Herman Melville in *Moby Dick*, a time when whalers burst out of the familiar Atlantic Ocean hunting grounds to pursue their prey into the Pacific and right around the globe. It was the age of the great sperm whale hunt when whalers slaughtered hundreds of thousands of the giant, toothed beasts so valued for their pure, odourless oil. It was also the period when New England shipowners and seamen emerged as the leading whalemen in the world.

I

The new era dawned early in 1789 when the 270-ton vessel *Emilia*, out of London, England, rounded Cape Horn and became the first whaleship to enter the Pacific. "On the success of our ship depends the Establishment of the Fishery in the

Pacific Whaling Grounds.

Jonathan Gladstone/j.b. geographics

South Pacific Ocean," the *Emilia*'s owner, Samuel Enderby, wrote, "as many owners had declared they shall wait till they hear whether our ship is likely to succeed there. If she is successful a large Branch of the Fishery will be carried on in those seas; if unsuccessful we shall pay for the knowledge."[1] No stranger to risk, the energetic Enderby had dispatched the first British whaleships to the South Atlantic grounds in the 1770s and was an outspoken voice of the London whale merchants in their dealings with the government. Along with his sons, he ran a company that Melville singled out in the pages of *Moby Dick* for special praise as the most vigorous of the London whaling houses.

The voyage of the *Emilia* ended a protracted struggle by Enderby to gain permission to send a ship into the Pacific. In the previous decade, the voyages of Captain Cook to Tahiti, Australia, New Zealand, Hawaii and the northwest coast of America focussed world attention on the Pacific as never before. Chroniclers of these voyages described an ocean teeming with whales. "There are a greater abundance of Whales and Seals rowling about these Straits, than I suppos'd were to be met with in any part of the World," Cook's second lieutenant, Charles Clerke, wrote about the waters off Tierra del Fuego, the doorway to the Pacific. "A fair account of them wou'd appear incredible."[2] But despite the prodigious surveys undertaken by Cook and other explorers, the Pacific remained largely innocent of outsiders, chiefly because the powerful British chartered monopolies, the East India and South Sea companies, refused to allow whalers into an area where they enjoyed unrivalled influence. Fiercely protective of their special privileges, the merchant companies wanted the whalers confined to the Atlantic where they could not interfere with the profitable trade to the Far East. But Enderby persevered and eventually won permission for specially licensed whaleships to enter the Pacific south of the equator as far west as 180 degrees longitude, an area encompassing the coast of South America and most of Polynesia. Jumping at this chance, he decided to outfit the 102-foot *Emilia* and send it on its way.

The *Emilia* faced risks beyond the usual ones associated with sailing in unfamiliar waters with no charts and only a second-hand knowledge of currents and landmarks. Captain James Shields also had to contend with the opposition of the Spanish.

South America was their domain, and they had shown a determination to keep whalers away from the coasts and out of the ports, even if it meant seizing ships and confiscating cargoes. It was an open question whether a whaleship could undertake such a long voyage into potentially hostile seas without secure harbours in which to resupply.

After visiting the Brazil Banks and the Falkland Islands in the South Atlantic, Captain Shields took the *Emilia* with its crew of twenty-one men around Cape Horn in January 1789. Fighting heavy weather and keeping a close lookout for hostile Spanish ships, the vessel sailed north away from the Horn in search of sperm whales. On March 1 the lookout at last spied a large pod and the boats took up the hunt, but the animals disappeared. Two days later, after a spirited chase, the honour of harpooning the first Pacific sperm whale belonged to the young first mate from Nantucket, Archaelus Hammond. Four other animals died that day, the first in the history of commercial whaling in the Pacific.[3]

Captain Shields continued north up the coast of Chile and Peru toward the equator, reluctant to venture too close to land for fear of running into a Spanish ship, but not liking to stray too

Bay of Islands. This New Zealand harbour was a favourite provisioning spot for Pacific whaleships.

far west into the unknown open ocean. He "spoke" only one other ship, a coasting vessel, the captain of which would not believe that Shields had sailed around the Horn from England. By mid-August the *Emilia* had collected 35,000 gallons of sperm whale oil and the crew was coming down with scurvy, so the captain turned and headed for home. Arriving in London in March 1790, he reported, "I never saw so many large Sperma Coeti Whales all the time I have been in the business as I have this voyage."[4] His words touched off a small stampede of whaleships toward the Pacific.

II

Prior to 1776 British shipowners left the so-called southern fishery to the Americans. Ships returning from the south did not even receive the bounty that had proved so successful at fostering an Arctic whaling industry from British ports. But the American Revolution presented new opportunities to the British. During the war, the American fleet was nearly wiped out. More than 85 percent of Nantucket's whaleships were captured or destroyed.[5] Several merchants transferred their vessels to English ports. Archaelus Hammond was only one among many island whalers who crossed the Atlantic to find work in the British fleet. American independence brought no relief. The British government imposed an import duty of over eighteen pounds sterling per ton on imported whale oil. The purpose of the duty was to stimulate the British industry, which it did, largely by denying the Americans access to what had been their major market. While Samuel Enderby and other British whalers rushed to fill the void, the New England whaling industry entered a period of crisis.

New England whalers faced a choice. They could try to survive on the restricted American market for oil and candles, a modest market that was glutted by oversupply already. Or they could seek shelter behind the British tariff wall by moving their operations, either to territory that still belonged to the British Empire and therefore enjoyed the protection of its preferential trade policies, or to the mother country itself. Nantucket, as the leading American whaling port, was at the centre of the debate.

During the Revolutionary War most Nantucketers showed little sympathy for the rebellious colonies. An exodus of whalemen began, which only increased after the conflict. Islanders

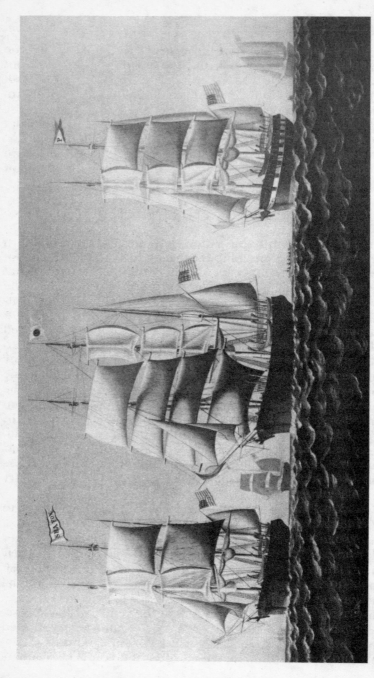

Gamming. Three American whaleships in the Pacific stop for a "gam," an exchange of news and gossip.

were loyal to their stern, Quaker principles and to their vocation as the world's foremost whalers. Loyalty to the new American nation meant little to them, especially if it interfered with their ability to conduct business. Many Nantucketers stood ready to move anywhere, as long as they were left alone to do their work, but as canny merchants with an appreciation of their own worth, they preferred to sell their services for a good price if they could.

Alternatives to emigration existed, and the Nantucketers explored them. They appealed to the Massachusetts Legislature to grant their island neutral status, hoping that this would convince the British to exempt Nantucket oil from the crippling import duty. In effect, they were asking to be excluded from the newly created United States, and not unnaturally the request was denied. Another alternative was to avoid the duty by shipping oil to a colonial port–Halifax, Nova Scotia, for example–where friendly merchants trans-shipped it to England as their own, duty-free. However, this subterfuge only took care of a small amount of Nantucket's oil. Islanders needed a better, permanent, solution, and relocation seemed to provide it.

In July 1785, Timothy Folger and Samuel Starbuck, two of Nantucket's leading whaling merchants, petitioned Nova Scotia's governor, John Parr, to allow them to establish a colony at Dartmouth, across the harbour from Halifax. Approval was quickly granted and the Nova Scotia Assembly promised the newcomers free land, money to build houses and assurances that their Quaker beliefs would be respected. Most importantly, the transplanted Nantucketers received British citizenship, which meant duty-free access for their whale products to the British market.

About forty families moved to Dartmouth from Nantucket, and soon the colony was sending out twenty-two vessels to the coast of Africa and the Brazil Banks. Then, as suddenly as they arrived in Nova Scotia, the Nantucketers departed. Whaling interests in London, who resented the assistance these upstart rivals were getting from colonial authorities, had opposed their presence in Nova Scotia from the beginning. As a result of their lobbying, Governor Parr received orders to stop encouraging the Dartmouth colony, and in 1792 the Nantucketers decamped for the Welsh port of Milford Haven, where generous arrangements were made to settle them.[6]

At the same time that the Dartmouth colony was getting on

71

its feet, another emissary left Nantucket to investigate the possibility of moving some of the island's whaling business across the Atlantic to England. William Rotch was the eldest son of the pre-eminent Nantucket whaling family of the late eighteenth century. A Quaker, related through his mother to the island's founding families, Rotch had been involved in the whale business most of his fifty years. Sober, shrewd, experienced, he was in many ways a natural choice to represent the besieged community. But Rotch was also an aggressive merchant whose attempts to control the whale-oil trade had won him enemies among the island's commercial élite. Not all Nantucket shipowners were as willing as Rotch to abandon their homes, no matter how bleak the future looked.

When Rotch arrived in London in July 1785, he found to his satisfaction that while "the Spirit of Whaling seems almost running to a degree of Madness" among London merchants, they were unable to find enough capable sailors to send out the number of ships they wanted.[7] Rotch offered to bring a colony of Nantucketers to live in Britain in return for financial assistance. Later the figure was set at about £20,000 for one hundred island families. But government officials were slow to respond and when an offer did come it was much less than Rotch wanted. Apparently the British were not as desperate for skilled whalers as they appeared to be. Rotch decided to take his proposal across the Channel to France to see if he could strike a better deal.

Before leaving London, Rotch met with Charles Jenkinson, Lord Hawkesbury, a veteran politician and the voice of the British whaling interests on the Board of Trade. The two men took a dislike to one another. "A greater enemy of America, I believe, could not be found in that body [the Board of Trade] nor hardly in the Nation," Rotch complained, a rather strange accusation from someone engaged in bargaining away his citizenship to the highest bidder.[8] His islander's pride was stung when Hawkesbury seemed to question the worth of investing in the Nantucket colony. "And what do you propose to give us in return for this outlay of money?" the British nobleman reportedly asked. "I will give you," Rotch responded haughtily, "some of the best blood of the island of Nantucket."[9] But Hawkesbury remained unmoved. He believed that he had the Nantucketers at a disadvantage and could bring them to England at a cheaper price.

For his part, Rotch refused to be made a fool of, and in the

Whaling capital. A painting of the New Bedford waterfront in the heyday of the whaling industry. This is the opening scene in a panorama depicting a whaling voyage painted on a roll of canvas a quarter of a mile long.

spring of 1786 crossed to France where in no time at all he came to an agreement. The French, anxious to acquire such a valuable addition to their own small whaling fleet, agreed to give the Americans free land and houses at Dunkirk, along with bounties on the oil they brought home. By 1789 the colony at Dunkirk, mainly a Rotch family operation, had a fleet of twenty ships out on the whaling grounds.[10]

But once again, the uncertainty of war and revolution clouded the future. As France was convulsed by the violence of the French Revolution and war with England drew near, William Rotch decided to pull his whaling vessels out of Dunkirk. The transplanted New Englanders were technically French, and once war broke out their ships were liable to be seized or sunk by British warships. In January 1793, William Rotch himself sailed from Dunkirk for London, two days before Louis XVI lost his head, two weeks before France opened hostilities against Britain. Once again the colony dispersed, some of the masters moving across the Channel to the Milford Haven settlement, others returning to New England where whaling was recovering from the blow dealt it by the Revolutionary War. William Rotch came home to Nantucket, but not for long. Though his welcome was cordial enough, some members of the community did not forgive his leaving the island, and he moved his business to New Bedford soon after.

III

While Yankee whalers worked to re-establish themselves, the extension of whaling into the Pacific continued. In the waters around Australia, whaleships doubled as transports carrying convicts to the newly-established colony at New South Wales. The second contingent of transports, arriving at Port Jackson in 1791, included another of Samuel Enderby's vessels, the *Britannia*. As the ship approached the coast of New South Wales, Captain Thomas Melville happily reported to his employer: "We saw Sperm Whales in great plenty. We sailed through different shoals of them from 12 o'clock in the day till sunset, all round the horizon, as far as I could see from the mast head. In fact I saw very great prospects in making our fishery upon this coast and establishing a fishery here."[11] After dumping his human cargo, Captain Melville went cruising after whales. Foul weather frustrated this voyage and the *Britannia* returned to port with only one whale, but the industry nevertheless was launched in the western Pacific.

A major stumbling block was the East Asia Company, which persisted in refusing to relax its restrictions on shipping in its chartered territories. London whale merchants convinced the government to pressure the company, however, and by 1798 the waters around southern Australia and New Zealand were open to whalers. One of the most active grounds was on the south shore of Tasmania, or Van Diemen's Land, as it was then called— the huge, green island off the southeastern tip of Australia. In August 1803, Lieutenant John Bowen led a small party of twenty-four convicts, eight soldiers and a handful of free settlers from Sydney to the mouth of the Derwent River to occupy Tasmania for the British and establish a convict colony there. No sooner had the new settlers arrived than they realized that they had stumbled upon an incredibly rich whale-hunting ground. Vast numbers of right whales visited the estuary of the Derwent from May to October every year to calve. "The river for some six weeks has been full of the whales called by the whalers, the Right or Black Whale," reported Lt.-Col. David Collins, Bowen's replacement as commander of the settlement. "Three or four ships might have lain at anchor and with ease filled all their casks. . . ."[12] Since one of the transports that carried the first convict settlers to the Derwent was a whaler, it did not take long for word of the discovery to spread. Ships anchored in the estuary while their boats went out among the surging black backs of the animals herding innocently in the river. By 1806 a shore station was in operation across from Hobart Town, the ragged settlement of huts and tents near the mouth of the river.

The Derwent was not the only place that the right whales swarmed each season. Migrating lazily northward from their feeding grounds near the Antarctic ice, the whales followed the shoreline of Australia and New Zealand, pausing in broad bays or at the mouths of large rivers. Wherever they went, the shore whalers followed, throwing up ramshackle shelters and warehouses and patrolling the coast in boats. When hunters killed an animal, they towed the carcass back to the station, hauled it up a ramp and stripped off its blubber, which was boiled in great trypots on the spot. At the end of the season, ships came to collect the catch and transport the men back to the settlements. In the heyday of shore whaling in the 1820s and 1830s, there were thirty-five stations in Tasmania alone, at least eighty in New Zealand and dozens along the south and west coast of Australia, wherever whalers found a sheltered harbour and supplies of fresh water and timber for fuel.[13] At many places along the coast, bay whalers also dropped anchor, not bothering to erect buildings on shore but instead using their ships as floating stations.

Deep-water whaling for sperm whales continued to expand at the same time that the right whalers were active. The first sperm whaler visited New Zealand in 1791, and by 1801 the governor could report that the industry was a qualified success. "Respecting the whale fishery on this coast, it has certainly succeeded so far that three Ships have gone home loaded with spermaceti oil. . . . Every ship that comes here, and indeed our Colonial vessels, always see great quantities of whales, but the main objection on the part of the whalers is the frequent gales of wind that happen on the coast."[14] But more than poor weather frustrated the New Zealand whaling industry. Just as important was the antagonism between the indigenous Maori population and foreign crews who put in at the islands for supplies. All too often outsiders tried to take advantage of the natives, with unhappy results.

In 1808 the trading vessel *Parramatta* stopped at the Bay of Islands, the principal harbour of New Zealand, on its way between Sydney and Tahiti. The local Maoris supplied the visitors with fresh food and water, but when they asked for payment the sailors tossed them overboard. As the *Parramatta* left the harbour a storm drove the ship onto land where a party of angry Maoris caught up with it and took their revenge by murdering everyone on board.[15] The following year the transport vessel *Boyd* was collecting spars at the northern port of Whangaroa when the natives attacked, killing everyone except a woman and three children who were rescued by another vessel. Once again the attack was motivated by revenge; a Maori chief who sailed as a crew member on the *Boyd* had been flogged by the captain on the way to New Zealand and apparently masterminded the bloody attack. The bloodshed did not stop there. A couple of months later a group of whaling captains at the Bay of Islands decided to hunt down the natives responsible for the *Boyd* murders. Launching an armed assault on a nearby island, ostensibly to look for captives and to disarm the natives, the whalers drove the Maoris onto the mainland, killing several and destroying their homes. In a letter describing these events, five of the captains warned "all commanders coming to this place to be constantly on their guard, the natives appearing determined and adequate to carry any single ship."[16]

While this familiar cycle of violence inhibited whalers in the western Pacific, vessels choosing to cruise farther east near South America were meeting opposition from the Spanish residents of the coastal regions. The continent was feeling the first stirrings of revolution against colonial rule and Spanish authorities feared that foreign whalers might aid the rebel cause. As a result, they harassed the whaleships and denied them permission to take on supplies in South American ports. Pacific whaleships were away from home for two years or more and their crews suffered horribly from scurvy when they had no fresh food.

Conflict with Spain was just part of the wider conflicts of the French revolutionary period in Europe. These wars wreaked havoc on the whaling industry. Slow-sailing and unarmed, whaleships were easy prey for hostile naval frigates. The risks of a voyage grew so great that many merchants withdrew their vessels from the business. In 1812, when Great Britain and the United States declared war, British whaleships armed themselves and carried two captains, the "whaling captain" and the "fighting captain." Setting off from Europe at the outbreak of hostilities, they managed to capture several American vessels that had been at sea and did not even know there was a war on. The British navy blockaded the coast of New England and intercepted whalers as they hurried home from distant voyages; in 1813–14 Nantucket lost half of its fleet this way.

On the west coast of South America, rebels in revolt against Spanish authority sided with Britain and used the war as a pretext to harass American whalers then in the Pacific. At length the Americans answered firmly. The United States government dispatched Captain David Porter and his 860-ton naval frigate *Essex* to the Pacific with orders to make life difficult for British shipping there. Learning that the Galapagos Isands were a favourite rendezvous for the whalers, Porter arrived at the archipelago in mid-April 1813 and began ambushing British ships. By the time the *Essex* was captured by a British vessel in Valparaiso harbour almost a year later, Porter had managed to put out of business one-half of the British Pacific whaling fleet, worth an estimated $2.5 million.[17]

The island of Nantucket was particularly devastated by the war. The whaling industry had recovered from the post-revolutionary depression and once again the island depended totally on whaling. As its ships surrendered to the British or sat idly at dockside unable to put to sea, the economy stagnated. By 1814 Nantucketers were finding it increasingly difficult to get food and fuel supplies from the mainland through the enemy blockade. Since the American Congress seemed unwilling, or unable, to help, desperate islanders approached the admiral of the British fleet with a request to lift the blockade. Seeing a chance to embarrass the United States, Admiral Sir Alexander Cochrane agreed and in August 1814 Nantucket formally renounced all support for the U.S. and agreed to terms of neutrality with the British. Once again islanders threw their patriotism overboard in an attempt to keep their whaling industry afloat. However, the war drew to a close before the implications of Nantucket's separate peace were revealed.[18]

IV

With the return of peace to the world's oceans in 1815, sperm whaling entered a long period of expansion. In the Atlantic the sperm whale population was decimated and ships extended their search for whales to the farthest corners of the southern oceans, spanning the Pacific and cruising through the Indian Ocean, along the shores of Arabia and on to the coast of India. By 1847, nine hundred ships and tens of thousands of men took part in this hunt worldwide. Alexanaer Starbuck estimated that between 1804 and 1876, whalers destroyed 225,521 sperms.[19] During the height of the fishery as many as 5,000 sperm whales were slaughtered each year in the Pacific alone. The sperm whale was not the only species captured—right whales still attracted interest, along with humpbacks and others—but the first half of the nineteenth century was the great age of the sperm whale fishery.

More than that, it was the age of the American sperm whale fishery. Between the American Revolution and the Napoleonic Wars, British whalers entered the southern fishery with enthusiasm, encouraged by high prices for oil, moderate subsidies, and the disarray of their American rivals. But during the 1820s, British interest in the southern grounds waned. The Pacific fleet had been devastated during the War of 1812. Long voyages lasting several seasons were very expensive. The price of oil in Britain tumbled as cheaper rapeseed oil replaced whale oil as a cleanser in the woollen textile industry and coal gas proved superior as an illuminant. At the same time, Australians began sending out their own whalers. It made little sense to send British ships halfway around the world when colonial oil could be imported more cheaply. In the early years of the 1820s the number of British ships in the Pacific dropped by 50 percent, then declined more slowly until by 1843 only nine southern whalers flew the Union Jack.[20] It was a sign of the times that late in the decade Charles Enderby, scion of the great British whaling family, left England never to return, choosing instead to base a new whaling venture in northern New Zealand.

The French briefly aspired to the place once occupied by the British whaling fleet. In 1816 the French government began offering bounties to ships going whaling. These premiums convinced several American captains to relocate on the Continent, particularly at Le Havre, which soon became the centre of the French industry. French ships kept pretty much to the South Atlantic where they patrolled in search of right whales, though in the 1830s they made the long voyage into the Pacific in increasing numbers. Still, the French fleet involved in the southern fishery in any one year seldom totalled more than forty ships, and after 1850 they ceased to be a presence on the whaling grounds.[21]

While European whaling nations lagged behind, the American fleet grew by leaps and bounds. By 1833, there were 392 American whalers at sea, more than all the other whaling nations combined. In 1846 the fleet reached its maximum size, 735 ships, about 80 percent of the whaleships in the world, worth more than $21 million and crewed by more than 18,000 sailors. Each year these ships brought home between 4 and 5 million gallons of sperm oil, another 6 to 10 million gallons of whale oil and anywhere from 1.6 million to 5.6 million pounds of whalebone.[22] American prices for all these products rose steadily through the period. Oil was still in demand as a lubricant in the rapidly industrializing economy, as well as an illuminant, and whalebone was used to make a host of products that required strength and flexibility. "Our whaling fleet may be said at this very day to whiten the Pacific Ocean with its canvass," observed the American naval commander Charles Wilkes in 1845 after returning from a cruise through the South Pacific, "and the proceeds of this fishery give comfort and happiness to many thousands of our citizens. The ramifications of the business extend to all branches of trade, are spread through the whole Union. . . ."[23]

This huge armada possessed an insatiable appetite for whales. As quickly as they depleted one whaling ground, the whales moved on in search of another. Following the arrival of the *Emilia* in the Pacific, the "In-Shore Grounds" along the coasts of Chile and Peru were hunted intensively. Some whalers travelled to Australia and New Zealand, while others ventured deeper into the central South Pacific.

One of these pioneers was the *Topaz*, a New England ship commanded by Captain Mayhew Folger. In February 1808, the *Topaz* approached the shores of Pitcairn Island and stumbled upon the answer to one of the most perplexing riddles in the history of Pacific navigation. Nineteen years earlier, thousands of miles to the east in the Tonga Islands, the crew of the British naval transport *Bounty* mutinied against its captain, William Bligh. The *Bounty* had just recently been in Tahiti waiting to collect a cargo of breadfruit seedlings for transport to the West Indies. During their lengthy stay the British seamen, used to the iron discipline of naval routine, grew fond of the island's relaxed way of life, and particularly of the local women with whom many formed liaisons. After setting adrift Captain Bligh and eighteen loyal crew members in an open boat, the mutineers returned in the *Bounty* to Tahiti where the majority decided to remain. A British man-of-war later rounded up these men and returned them to England for trial. Three eventually hanged. The rest, led by the ringleader of the mutiny, Lieutenant Fletcher Christian, merely paused at Tahiti to collect their native women, then set sail and disappeared to the west.

This was as much as was known when the *Topaz* drew near to Pitcairn Island, believed at the time to be uninhabited. Captain Folger was amazed, therefore, when he saw a boat paddled by three natives coming out from the island. "On approaching her," he wrote in his log, "they hailed in the English language, asked who was Captain of the ship and offered me a number of cocoanuts which they had brought off as a present, and requested I would land, there being as they said a white man on shore." Folger continued with his story: "I went on shore and found there an Englishman by the name of Alexander Smith the only person remaining out of nine that escaped on board the ship *Bounty* under the command of that arch mutineer, Christian." Smith told Folger that after leaving Tahiti with their "wives" and six other natives, the mutineers had sailed the *Bounty* to Pitcairn where they destroyed the ship and settled down to live. According to Smith, one of the Europeans went mad and drowned himself and another died of disease. Then, for some reason, after about four years the Tahitian men in the colony rose up against the Europeans and slaughtered them all. All except Smith, who with the help of the women managed to kill the rebels, leaving himself, eight or nine women and several small children as the sole survivors of the bloody episode. From this nucleus the colony began again, and by the time Folger found them they were living quite comfortably.[24] Years later, when a British naval frigate visited Pitcairn Island, Smith, whose original name turned out to be John Adams, was still alive, but so much time had passed since the mutiny that the British left its only survivor in peace. He died on March 5, 1829.

That the fate of the *Bounty* mutineers remained a secret for so long indicates how few ships travelled through the South Pacific. However, this did not long remain the case. In 1818, as sperm whales became harder to find close to South America, Captain George Gardner in the whaleship *Globe* discovered a new ground about 1,000 miles off the coast of Peru in mid-Pacific. Within two years more than fifty ships were visiting this "Off-Shore Grounds." But already other American vessels had reached the Hawaiian Islands in 1819 and the waters southeast of Japan the next year.

By the mid-1820s whaleships had settled into fairly regular sailing patterns, travelling systematically from place to place depending on the season and what they heard about the luck other whalers were having on the different grounds. After leaving home, usually late in the spring, New England whalers crossed the Atlantic to the Cape Verde Islands and the Azores, the so-called Western Islands, where they hunted whales and put in for fruit, vegetables, fresh meat and water. Riding the prevailing winds south back across the Atlantic, they sailed through the Brazil Banks and around Cape Horn into the Pacific. As they moved north again up the coast of Chile, the snow-capped spine of the Andes was visible behind the low shore in the distance. "In the bright sun," Benjamin Doane wrote during his 1846 cruise, "the yellow sands which covered the mountain sides shone like running rivers of gold." Off Caldera in northern Chile one of the veteran crewmen on board might bring up the story of "Caldera Dick," the legendary sperm whale that for many years patrolled this stretch of coastline. Several times the giant animal was harpooned but always it turned on its attackers, smashing their boats to pieces with its broad flukes or crushing them in its jaws. The story went that a wily Nantucket captain managed to make fast to Caldera Dick with a large wooden cask, tightly bound by metal hoops. The whale could not rid itself of the annoying cask, which acted as a powerful drag every time the animal sounded, nor could it manage to smash it. At length the whale lay exhausted on the surface and the whalers moved in for an easy kill.

The whaleships stopped at a number of ports to resupply on their way up the coast. These way-stations included Juan Fernandez Island, famous as the spot where, from 1705 to 1709, the marooned British seaman Alexander Selkirk lived alone for fifty-two months, inspiring Daniel Defoe's famous novel, *Robinson Crusoe*. Valparaiso, Talcahuano, Callao, Payta and Tumbez also welcomed whalers with fresh water, produce, meat and squalid wineshops and brothels where the sailors passed their shore leave. Eventually the ships reached the Galapagos Islands, six hundred miles off the coast of Ecuador. This barren, volcanic archipelago provided giant tortoises, which the whalers captured by the hundreds to make a welcome change in their diet. "They neither eat nor drink, nor is the least pains taken with them," explained whaler Owen Chase. "They are strewed over the deck, thrown under foot, or packed away in the hold, as it suits convenience. They will live upwards of a year without food or water, but soon die in a cold climate."[25] So popular were these animals with whalers and sealers who visited the islands that today three of the fourteen subspecies are extinct and the original population of 250,000 now numbers only about 15,000.[26] The other attraction of the Galapagos was a visit to Post Office Bay on Charles, or Floreana, Island. Near an emergency cache of supplies in case of a wreck, someone had nailed up an oil cask to serve as a mailbox, and seamen routinely dropped off letters for home or messages for other ships in the fleet. Incoming whalers looked through the "mail" to see if there was any news for them; outgoing ships checked to see if there were any messages to carry home.

Whalers endeavoured to refit at one of the South American ports and be out on the Off-Shore Grounds by November when the sperm whales were most plentiful. After a cruise of two or three months they either sailed north to the Hawaiian Islands or set a more westerly course along the equator into Polynesia, where once again they put in at friendly harbours to take on supplies and scrape the barnacles and weeds from their hulls. Every island group had a port of call that welcomed the whalers for the business they brought. In the Marquesas it was Resolution Bay on the island of Nukahiva. "But few ports in the Pacific offer greater facilities for refreshment of ships than Resolution Bay," remarked Frederick D. Bennett after his visit in the 1830s. "Supplies of hogs, poultry, and vegetables are abundant, of excellent quality and may be obtained on very easy terms. Wood and water are equally available."[27] At the same time, 150 whaleships a year were dropping in at Papeete on Tahiti, and others visited harbours in Fiji, Samoa and Tonga. To the south, in New Zealand, the Bay of Islands was the most important supply centre in the Pacific until about 1840, when it was replaced by the harbours of Hawaii.

Continuing on their way, members of the whaling fleet hoped to arrive on the Japan Grounds, roughly located between

82

the east coast of Japan and the Bonin Islands, in about May. Sperm whales were plentiful right across the Pacific on either side of the thirtieth parallel of latitude through the summer, after which the ships paused at the Hawaiian ports—Honolulu or Lahaina—then returned south to resume whaling on the Off-Shore Grounds in November when the whole cycle began once again. The objective was a full ship, a hold bursting with casks of oil, and Pacific whaling captains gave themselves anywhere from two and a half to four and a half years to achieve it.

As an alternative to this itinerary, whalers might choose to visit the Indian Ocean, which increasing numbers of them did after 1815. By 1845 there were ninety American and three dozen British vessels cruising through the Mozambique Channel and around Zanzibar, the Seychelles, Mauritius and Madagascar.[28] Briefly, sperm whales were plentiful near Ceylon, and after 1850 vessels sailed as far as the coast of Arabia. Whaling in the Indian Ocean peaked in the late 1840s, then began a steady decline. Many ships rounding the bottom of Africa were not diverted by prospects in the Indian Ocean at all. They were heading without pause for Australia where they liked to cruise the Middle Ground between New South Wales and New Zealand in March, April and May until the boisterous weather of a southern winter came on. Then they might head east, planning to cross the Pacific in time to be on the Off-Shore Ground in November, or veer north into Polynesia until calmer weather called them back south.

All of these different routes depended on many vari-ables—currents, weather, the success each captain was having, the information picked up as the fleet moved around in search of whales. The important point is that whalers did not wander aimlessly across the face of the globe hoping to come across their prey. On the contrary, they understood that the animals followed broad migration patterns linked obscurely to the currents and the seasons. At certain times of the year, whales were more likely to be found at certain places than others. It was the captain's job to know, not to guess, where these places were and to see that his ship arrived there safely, ready for the chase.

V

The Pacific was full of hazards for the nomadic whalers. As Alexander Starbuck pointed out, "The sea was comparatively unknown; what charts there were in existence were full of inaccuracies, and the first intimation that many a vessel had that she was sailing on dangerous ground was the splash of the breakers close at hand, or the grinding of her keel upon the treacherous rocks."[29] Malevolent reefs surrounded the Pacific islands and the sea between was sprinkled with uncharted shoals and atolls. As a result, shipwrecks were common, leaving survivors adrift on the open ocean with only the food they could carry and no shelter from the torrid sun and heavy gales. The captain's wife who managed to escape a shipwreck and survive on a deserted island by eating the rats that had swum ashore along with her was only an extreme example of the fate that haunted the seaman's imagination.[30] Almost as harrowing was the story of the crew of the New Bedford whaler *Canton*, which sank after running onto a reef near the equator in March 1854. Getting away in boats, the sailors rowed to a small island where they lived for four weeks on food they managed to rescue from the wreck. When supplies ran low, they put to sea again and drifted for nearly two months, subsisting on meagre rations of water and biscuit. When they at length arrived at one of the Marianas Islands, the governor suspected they were pirates and did not let them land. He did, however, send out supplies to the boats, and after another six weeks at sea the desperate flotilla found safety at Guam.[31]

Storms, of course, threatened Pacific whalers, as they did mariners everywhere. The area around New Zealand was especially unsettled, though heavy seas and strong gales could strike anywhere. Few storm stories could rival the case of the Nantucket ship *Henry*, returning in July 1813 from a long cruise in the Pacific. Just a few days from home the vessel foundered in a fierce storm. Only the bowsprit remained above water and the crew clung to it for their lives. For forty days the sailors dangled from the wreckage, surviving on bits of bread and water retrieved from the hold, scanning the horizon for a rescuing sail. One by one they weakened, dropped off and disappeared in the grey seas. By the time rescue came, only five men were alive.[32]

But the greatest danger for the whalemen was the whale itself. It was not unusual for right whales and bull sperms, when pursued, to turn on their tormentors, crushing boats in their gaping jaws or smashing them to matchsticks with their powerful flukes. Survivors of these attacks clung to the wreckage until another boat plucked them from the sea, but if the chase carried them out of sight of their comrades, or if night fell, or if poor weather set in, they might easily perish in the cold water before rescuers got to them.

The most famous of these homicidal whales was "Mocha Dick," an immense bull sperm with a ragged white scar across its forehead. "Mocha Dick" began his war on Pacific whalers about 1810 when he first attacked a whaleboat off the coast of Chile. For the next fifty years he eluded capture, smashing numberless boats that came after him, killing at least thirty men in the process. Aged and blind in one eye, he finally succumbed to a harpoon in 1859. When flensers began to cut in they found nineteen old irons embedded in his body. No wonder Mocha Dick became the model for the great white whale of fiction, Moby Dick.

Alexander Starbuck related another unusual story of a whale that attacked and destroyed a boat after being harpooned. While the ship's captain, who happened to be the header in the boat, paddled about in search of some floating object to grab hold of, the whale reappeared. As if it knew who was responsible for its troubles, the animal repeatedly raised its massive head out of the water and dropped it on the captain, driving him underwater. "Thus for three-quarters of an hour that whale and I were fighting," recalled the captain. "The act of breathing became laboured and painful; my head and shoulders were sore from bruises, and my legs had been pounded by his flukes; but it was not until I found myself swimming with my arms alone and that my legs were hanging paralyzed, that I felt actually scared." At this point, the battered sailor lost consciousness, but as he lay in the water another boat at last got close enough to rescue him. "I was put to bed, a mass of bruised flesh," he concluded. "It was several weeks before I was able to take my place in the head of my boat again."[33]

If vengeful whales routinely smashed boats, they occasionally also attacked the ships. Instances were few, but dramatic. In October 1807, a large sperm whale in the Atlantic struck the Nantucket vessel, *Union*, smashing a hole in its side. As the vessel settled in the water, the crew took to the boats and after a week at sea managed to reach the Azores.[34] In August 1851 another sperm whale, this time near the equator in the Pacific, rammed the *Ann Alexander* after first chewing up two of its boats. The collision burst the hull below the water line and the ship went down. Again the crew got away in the remaining boats. Five months later a New Bedford vessel killed this particular whale; it was identified from the two harpoons in its back with markings from the *Ann Alexander* and the pieces of ship's timber embedded in its forehead.[35]

Like the barnacled hide of an old right whale, the history of whaling is encrusted with such stories. But the most dramatic incident, without which America might have been denied one of its great works of literature, involved the whaleship *Essex*, cruising the mid-Pacific in November 1820. The *Essex* was already fifteen months out of Nantucket with eight hundred barrels of oil in its hold when early on the morning of November 20 the lookout sighted a group of sperm whales and the boats flew to the attack. First mate Owen Chase, whose narrative is the main source for the tragic events that followed,[36] planted his harpoon in the back of a large bull sperm, but the oarsmen were not quick enough and the angry animal burst a hole in the boat with its flukes. Chase cut the whale loose and by stuffing the hole with their jackets and bailing frantically the sailors managed to stay afloat long enough to return to the *Essex* for repairs.

Two other boats remained in the hunt and Chase was planning to patch his own quickly and get back out on the water when he noticed a large whale, perhaps the one he had earlier harpooned, surface quietly about a hundred yards away. As he watched, the whale began swimming toward the ship, picking up speed as it approached. Chase shouted to the sailor at the helm to change course, but it was too late. The animal struck the *Essex* near the bow, sending a shudder through the vessel, bringing it to a sudden halt.

The stunned whale moved slowly away, then began to convulse, thrashing about in the water and lunging with its huge jaws. On board, Chase believed the ship was sinking and was getting ready to abandon it when someone called out that the whale was coming back. "I turned around," recalled Chase, "and saw him about 100 rods [550 yards] directly ahead of us, coming down apparently with twice his ordinary speed, and to me at that moment, it appeared with tenfold fury and vengeance in his aspect. The surf flew in all directions about him, and his course towards us was marked by a white foam of a rod in width, which he made with the continual violent thrashing of his tail; his head was about half out of water, and in that way he came upon, and again struck the ship.''

The bow of the *Essex* was now thoroughly crushed and filling with water. Sailors only had time to load a spare boat with a few navigation instruments and shove off before the 238-ton vessel fell over on its side and settled in the water. The men were speechless with shock and despair. The whole episode took less than ten minutes. In that time they had lost food, shelter and all possessions but the clothes on their backs. They were stranded a thousand miles from the nearest land with only three fragile whaleboats to carry them across a storm-tossed ocean to safety.

Meanwhile, the other two boats, unable to see their ship but completely unaware of what had happened, came looking for their compatriots. Distress quickly gave way to the practical needs of rescuing supplies from the wreck. Smashing through the planked hull of the derelict the men managed to retrieve limited supplies of hard bread and water, some tools and a few turtles. As night closed in and the boats rose and fell on the swell next to the wreck, each of the twenty crewmen contemplated the desperate situation. "To shed tears was indeed altogether unavailing, and withal unmanly," wrote Chase, "yet I was not able to deny myself the relief they served to afford me." The mate was convinced that the whale had fully intended to ram the ship out of vengeance for the attack on its pod. "It is certainly . . . a hitherto unheard of circumstance, and constitutes, perhaps, the most extraordinary one in the annals of the fishery."

After spending another day installing masts, making sails and building up the gunwales of the three boats with stray boards, the tiny flotilla set sail for the southeast, heading for the coast of South America. Six men were in Owen Chase's boat; seven occupied Captain George Pollard's and seven more the second mate's. All hoped that another whaleship would shortly pick them up. Meanwhile, they survived on meagre rations of bread and water. At first the weather was stormy, pummelling the boats with high winds and heavy rains that threatened to swamp them. Then in mid-December the winds dropped away to nothing, and a pitiless sun glared down from a cloudless blue sky. The men sought protection by crawling under the sails in the bottom of the boat. The daily ration was reduced by half; thirst was perpetual. The men fell to drinking their own urine. One day four flying fish crashed against the sail and tumbled into Chase's boat. Immediately the famished sailors pounced on them, devouring head, wings, scales and all without even bothering to kill the fish. "Our sufferings during these calm days almost exceeded human belief," recalled Owen Chase.

On December 20 the trio of boats reached Henderson Island, an unoccupied island in the Pitcairn group about halfway between Easter Island and Tahiti. At first they thought they could survive until rescue came. But it became evident that there was not enough food on the island to keep them alive and they set sail again, leaving behind three of the crew who preferred to take their chances on dry land. The *Essex* survivors were unaware that only a day's sail to the west lay inhabited Pitcairn Island. Instead the boats headed southeast into the open ocean. Quickly the good effects of their stay on shore wore off and the sailors once again were reduced by hunger and thirst. On January 10, 1821, the first man died. The others sewed up his body in his clothes, attached a rock to his feet and dropped him over the side of Owen Chase's boat.

During a fierce storm, Chase's group separated from the others and he and his four companions continued on alone. By the end of January another sailor had died and the survivors, too weak to hold a course, let the boat drift. On February 8 only three men were left as Isaac Cole died a slow, painful death. The others were preparing to toss the body overboard when Owen Chase raised another possibility. They could eat it. Without much discussion, his crew agreed and fell to butchering Cole's body. First they ate the heart, then hung some of the meat in thin strips to dry in the sun while roasting the rest in a turtleshell for future eating. The next day some of the drying flesh had spoiled so all the rest was roasted. It lasted six or seven days and restored the sailors enough that they could resume steering their boat.

Finally, on February 18 at seven o'clock in the morning, ninety-one days after the sinking of the *Essex*, one of the survivors staring aimlessly at the horizon sighted a sail. It was the English brig, *Indian*, and when it came up beside the boat its crew looked down on three emaciated figures lying in the bottom, their skin covered in large ulcers, their limbs swollen and painful, their lips cracked and bleeding. "We must have formed at that moment, in the eyes of the captain and his crew, a most deplorable and affecting picture of suffering and misery," recalled Owen Chase. "Our cadaverous countenances, sunken eyes, and bones just starting through the skin, with the ragged remnants of clothes stuck about our sun-burnt bodies, must have produced an appearance to him affecting and revolting in the highest degree."

The *Indian* carried Chase and the two others to Valparaiso where within three weeks another whaler arrived with two survivors from Captain Pollard's boat. These men, too, had lived through months of hellish suffering. When their food ran out early in February they drew lots, the loser agreeing to give up his life so that the others might eat his flesh. As it turned out it was young Owen Coffin, Pollard's own cousin, who pulled short straw. The captain offered to spare the boy. "My lad! My lad! If you don't like your lot I will shoot the first man who touches you," he records himself as saying. But Coffin was past caring, or perhaps he preferred to die with a show of courage. Charles Ramsdell drew short straw on the second draw and it was he who shot the boy. Later, another man died of exposure, leaving only Pollard and Ramsdell to be rescued.

The seven men in the third boat were never seen again; a merchant vessel picked up the three who remained on Henderson Island and landed them in Australia.

But the story of the *Essex* does not end with the rescue of its survivors. First mate Owen Chase returned to Nantucket in June 1821, and within four months had published, with the assistance of a ghostwriter, his "Narrative of the Most Extraordinary and Distressing Shipwreck of the Whale-Ship Essex." By the end of the year, however, he was back aboard a whaleship where he spent most of the rest of his life. (Not so Captain Pollard. On his very next voyage his ship struck an uncharted reef and sank. Pollard returned to Nantucket safely, but no one entrusted him with another command. He ended his days as a night watchman.)

Meanwhile, in 1841, a young Herman Melville was sailing as a forecastle hand on the Pacific sperm whaler *Acushnet* when his ship paused to exchange news with another from New England. While visiting the other ship Melville fell into conversation with

William Henry Chase, the son of Owen Chase, who loaned the
future novelist a copy of his father's "Narrative." The story of a
great whale sinking the ship that attacked it deeply impressed
Melville and later, when he had returned from his adventures in
the South Seas and was writing *Moby Dick*, he kept his own
well-thumbed copy of Chase's volume open in front of him.

THE AMERICAN WHALEMAN

It would be better to be painted black and sold to a southern planter than be doomed to the forecastle of a whaling ship.

–an American whaleman

The ports of New England were home to the American whaling armada. From Long Island's Sag Harbor in the south to the coast of Maine in the north, these industrious seaside towns built economies that relied on whaling for their lifeblood. Supplies of food and clothing for the voyage piled up in warehouses, ready for the fleet's departure. Ships were cleaned and painted, their leaky seams recaulked, their frayed rigging repaired or replaced. Sailmakers prepared the lengths of white canvas, coopers assembled the hundreds of wooden casks and barrels to hold the cargo of oil, blacksmiths forged the harpoons, lances and flensing spades. According to one estimate in 1833, for every seaman who shipped aboard a whaler, six landlubbers made their living from the whale business.[1]

During the so-called golden age (roughly 1815 to 1860), New Bedford emerged to replace Nantucket as the most important whaling port. Founded in 1755, New Bedford surrounds a sheltered harbour at the mouth of the Acushnet River on the south coast of Massachusetts. The town's founder, Joseph Russell, owned a few whaleships, but it was the arrival of a branch of the Rotch family firm from Nantucket around 1770 that launched New Bedford into whaling. During the Revolution, British raiders torched the waterfront and destroyed thirty-four whaleships, but the town rallied and by 1805 it was the second-busiest port in New England and gaining on the leader, Nantucket.

For a hundred years Nantucket had sent its ships away on sperm whaling voyages. They had pioneered many new grounds, survived storm and war, to bring home the oil that was the island's staple export. But as voyages became longer and ships bulkier, Nantucket harbour revealed a serious handicap. A sandbank extending across the entrance made entry difficult for large ships. Island merchants tried to get around this problem with the help of a cumbersome apparatus known as a camel. Resembling a floating wooden pen, the camel was towed out to the whaleship waiting beyond the bar. Once it was placed snugly around the ship, water was pumped out of the camel so that it, and the vessel it embraced, became more buoyant and able to cross the shoal into the harbour. The use of a camel was time-consuming, and difficult in rough weather; it never did completely solve the problem.

Equally as important as the inadequacy of their harbour was the reluctance of Nantucket whale merchants to diversify. Sperm whalers they had been since Captain Hussey made his pioneering voyage and sperm whalers they remained, even as the giant animals grew scarcer and fleets from rival ports turned their attention to other, more profitable species. Gradually, Nantucket's fleet dwindled, until in June 1870 the brig *Eunice H. Adams* arrived home with its cargo of oil, the last whaler to cross the bar into the island's harbour. New Bedford, on the other hand, had a capacious harbour and merchant owners who sent their ships after any type of whale that would turn a profit. By the 1840s the mainland village sent out over 200 ships, one-third of the entire American fleet. And by 1857 fully one-half of all Yankee whalers, 329 vessels, sailed from New Bedford.[2]

I

Seamen aboard their sleek merchant ships liked to ridicule the whaleships for their heavy slowness and squat, ungainly appearance. Unquestionably, the "blubber hunters" were built for strength, not speed—broad of beam to handle a 60-ton whale carcass hanging alongside, with masts and rigging strong enough

Whale attacking a boat. Sometimes wounded whales turned on their attackers, crushing the fragile cedar boat between enormous jaws.

to stand the terrible strain of hoisting the bulky strips of oil-soaked blubber on board. As voyages extended from a few months to several years, the ships grew larger to accommodate the supplies and cargo. The first vessels to reach the South Atlantic were between 100 and 180 tons in size; the *Emilia*, first whaleship in the Pacific, was 270 tons; and by the 1820s the fleet was approaching an average of about 350 tons.

As ships grew larger, the crews that sailed them by all accounts grew more motley. In the early days of American whaling, New England ships were crewed by New England men. Growing up, young boys inhaled the atmosphere of a whaling port and thought no other profession was worth their attention. They were, explained Owen Chase, "captivated with the tough stories of the elder seamen, and seduced, as well by the natural desire of seeing foreign countries, as by the hopes of gain. . . ."[3] Sons followed fathers into the business and expected their sons to follow them. Few got wealthy, but whaling offered pretty well the only source of social prestige in the New England seaport towns. No young woman, it was said, paid a moment's notice to the attentions of a suitor who was not a whaler.

Indians and blacks were exceptions to the homogeneity of

the early whaling crews. Indians in New England engaged in boat whaling from the shore alongside the early settlers. On Nantucket, whites filled the positions of harpooners and steerers in the boats of the shore fishery, but most of the oarsmen were local Indians, the proletariat of the industry.[4] The native population of the island declined rapidly until an epidemic in 1763 virtually wiped out the few remaining Indians. By this time owners were recruiting "coofs," off-islanders, to fill their crews. Some of these were natives from Gay Head on Martha's Vineyard and Montauk Point on Long Island; by 1820 it was a rare whaleship that did not have two or three among its crew. Blacks entered the

A young Ishmael. In 1856, 18-year-old James McKenzie has just returned from a Pacific whaling voyage. Six years later he disappeared after being washed overboard in a gale.

fishery following the Revolution when the Nantucket fleet was getting re-established and needed increasing numbers of crewmen. Living on the island in their own community of shacks, called New Guinea, blacks came to fill about a quarter of the positions on a New England whaleship.

In 1808 a Baltimore newspaper described Nantucket whalemen: "Indefatigable, and inaccessible to fear, they navigate during whole years in almost open boats, exposed to all the solitude and perils of the sea."[5] This was the romantic view. The reality was becoming something else. With the rapid expansion of whaling after 1815, New England seaports failed to supply enough sailors from their own populations and crews came increasingly from the wider world. Recruiters rounded up young men all down the eastern seaboard, from the forests of Vermont and New Hampshire, from towns and villages across the Midwest and ports around the Great Lakes. These newcomers had no experience of the whaling business. They were inland boys with a naïve taste for adventure, lads like Melville's Ishmael, "tormented with an everlasting itch for things remote." They were also cursed with an inability to see through the promises of unscrupulous recruiting agents who turned their heads with truthless descriptions of the joys of life at sea. Shepherded by train and wagon to the whaling ports, green hands were lodged in cheap boarding houses and supplied with liquor and women to keep second thoughts at bay until their ship departed.

But soon, even America could not supply the appetite of its growing fleet for labour, and crews took on an international character. The typical whaler left port with a crew of about thirty: the captain, four mates, four harpooners, a cooper, a blacksmith, a cook, a steward, a cabin boy and the rest regular seamen. Not all these positions were filled at the start of the cruise. Captains usually hired the remainder at the Azores and Cape Verde Islands, the first stops on the long voyage. The residents of these Portuguese possessions eagerly went to sea aboard American whaleships to escape conscription into the Portuguese army and, in the case of the Azores especially, to escape the grinding poverty and famine that afflicted their native islands. By the 1850s about 3 percent of the labour force in the American fleet came from the "Western Islands," a number that increased to at least one-quarter during the following decades. Islanders earned very poor pay, even by the standards of the whaling business. Azoreans, because they were lighter skinned,

Nantucket sleighride. Rising winds and heavy seas make it impossible for this boat to keep up the chase and the steerer is about to cut the whale free. At the rear of the boat, a sailor is keeping the line wet so it does not burst into flame. The painting is by J. S. Ryder.

received marginally better treatment, but the black-skinned Cape Verdeans were at the bottom of the shipboard social ladder, right next to the American black.[6]

As a ship made its way from port to port through the South Pacific, it lost members of the crew to desertion, sickness, injury and death, and replaced them as often as not with escaped convicts from Australia, Kanakas from Hawaii, Maoris from New Zealand and natives from many other South Sea islands. Once they had overcome an original shyness, natives of the South Pacific were eager to explore the world beyond their tiny islands and happily signed on as crew. Captains prized the islanders for their incredible eyesight, which allowed them to spot whales at a great distance, and for their boldness once the chase began. "The mixture of people to be found amongst the South-Seamen is extraordinary," remarked one eyewitness in the 1830s. "Not one of the islands of the Pacific and South Seas, but furnish many useful hands for the fisheries." He then proceeded to list the origins of a typical crew: "Chilians, Peruvians, Patagonians, every cross of the Spaniard, Portuguese, and South American Indians, as well as some Malays, and samples of every European race, Dane, Swede, Dutchman, Frenchman, Italian, all are here mixed up. Negroes who have been emancipated or purchased their freedom, compose, in some instances, a considerable portion of the crew of Yankee whalers, and numerous convicts who have in various ways escaped from the colonies, swell the number of this varied and oddly assorted throng."[7]

The changing composition of the crew had important ramifications for life aboard American whaleships. Captains and mates were still predominantly New Englanders for whom whaling was a profession. But most of the regular hands were young, inexperienced and not suited to the fishery by any inclination or skill. One historian of the industry estimated that "green hands" comprised one-half of the number of crewmen shipped annually.[8] When Benjamin Doane sailed as a harpooner aboard the Pacific whaler *Athol* in 1845, he described the crew as being "the usual nondescript lot always found on whaleships, composed of everything but sailors. Some of them had never been to sea before. One was an ex-college professor; another had been a clerk in a bank. They had been ruined by drink. A third had been educated for the ministry. . . ."[9] Though merchant owners sometimes preferred to make a crew out of such unpromising material, expecting they would accept lower wages and inferior living conditions, the result was a steady deterioration in morale aboard ship, and an increase in the use of strong discipline and in the number of desertions.

II

Whaling crews were exploited by owners and agents. "There is no class of men in the world who are so unfairly dealt with, so oppressed, so degraded, as the seamen who man the vessels engaged in the American whale fishery," reported J. Ross Browne, a young crewman, after returning from a cruise to the Indian Ocean in the 1840s. Browne wrote his account expressly to draw public attention to the sorry plight of the whaling sailor who, he claimed, "is subject to severe labour, the poorest and meanest fare, and such treatment as an ignorant and tyrannical master, standing in no fear of the law, chooses to inflict upon him."[10] Elmo Paul Hohman, one of the few historians to study the working life of the American whaleman during the so-called golden age, agreed that it "was as hard and cheerless as that of any group of free workers in the history of the United States."[11] Sailors were systematically cheated by outfitters, poorly paid and badly cared for by owners, and tyrannized by captains. American whaling, wrote Hohman, "throughout the period of its dominance and decline, had all the essential characteristics of a sweated industry."[12]

Each member of the crew received wages in the form of a percentage of the net value of the cargo. This amount, called a lay, was not paid until the end of the voyage when the size of the cargo was known. In the meantime, during the many months they were at sea, sailors relied for spending money on their savings, which few had, and on advances that would later be deducted from their pay. The size of each sailor's lay depended on many things: his skill and experience, the availability of men to choose from, the price of whale products, wage rates ashore and in other branches of the maritime trades. Generally, an experienced captain received a lay of one-eighth to one-tenth; that is, one-eighth to one-tenth of the value of the cargo the ship brought home. Mates, harpooners, and coopers received up to one part per hundred, ordinary seamen from one part per hundred to one per one hundred sixty, and "green hands" and cabin boys no more than one part per two hundred fifty.

The reasoning behind this unusual form of payment was

Flensing a whale. Whalers climbed onto the back of the floating carcass to cut away its coat of blubber with long-handled cutting spades.

that sailors worked harder and took greater risks if they had an immediate interest in the success of the voyage. However, the profits from a voyage depended more on the whims of the whale-oil market than the willingness of crewmen to chance the perils of the hunt. In practice, the system of lays shifted part of the risk of a whaling voyage from the owner of the whaleship to its crew. If the vessel failed to return with a profitable catch, or if the price of whale products dropped, the crew received almost nothing at the end of the long voyage, and the owner would have successfully reduced a major expense.

Hohman estimated that in the mid-1800s an ordinary seaman on a whaling voyage earned between three and eight dollars a month, an amount "decidedly lower than the average wages paid in the merchant marine and in unskilled occupations on shore."[13] And even that pittance exceeded what most received since a bewildering range of expenses was deducted from the lay. Before the ship sailed, a sailor outfitted himself with clothing and other necessaries for the voyage, bought at inflated prices from an outfitter's store. Often the outfitter was also manufacturer, hiring local people to produce clothing and other items for the sailors. Hohman, who had little good to say about the outfitters, described the system: "These shrewd tradesmen succeeded in building up a system in which they bought the coarsest raw materials, had them made into poorly-fashioned articles by cheap labour, and then sold the resulting low-cost goods to economically helpless customers at exorbitant prices."[14]

Most departing seamen had no money to pay for these goods. The money was advanced them against the proceeds of the voyage and while they sailed distant seas this loan collected interest at an average rate of 25 percent. In other words, outfitting charges of $100, not at all unusual, amounted to $125 by the time the sailor returned to pay them. Meanwhile, he received no interest on the wage that was accumulating for him, but that was not paid until the voyage was over. During a voyage, members of the crew purchased tobacco, clothing, needle and thread, blankets, knives and the like from the ship's store, called the slop chest. Once again, prices were high and purchases were paid for as advances against wages. When sailors wanted money for spending at ports of call they did not receive cash. Instead they purchased goods on credit from the store at inflated prices, then resold them for less on shore. The captain heartily

"Life in the Forecastle." A depiction of off-duty sailors in their cramped quarters aboard a whaleship.

approved this formalized extortion, for it was his privilege to keep some or all of the profits from the slop chest!

As well as these various cash advances, crew members customarily paid small fees to cover the costs of preparing the ship for sea, unloading the cargo and cleaning the vessel once it returned to port. All these charges and advances, noted meticulously in the ledger, came out of the lay of each man, and sailors habitually returned from long whaling voyages to discover that they owed their employer money rather than vice versa.

One thing for which seamen did not pay extra was their berth and meals aboard ship. However, by most accounts, these items were not worth much anyway. While the captain and his mates lived fairly comfortably in their staterooms toward the stern of the vessel, and the steward, tradesmen and harpooners lived in steerage amidships, the rest of the crew, the ordinary seamen, inhabited the forecastle. In this low, dark cubbyhole between decks near the bow of the ship the men slept in hard berths ranged in tiers along the walls. Sailors' chests and bags cluttered the greasy floor. A hatch in the deck provided the only entrance, the only source of daylight, and it was shut during heavy seas. This cramped compartment housed fifteen to twenty men ("a number at once tragic and ridiculous," remarked Hohman) and as the ship roamed the torrid southern latitudes the atmosphere below decks became suffocating. "The forecastle was black and slimy with filth," recalled J. Ross Browne of his 1842 cruise, "very small, and as hot as an oven. It was filled with a compound of foul air, smoke, sea-chests, soap-kegs, greasy pans, tainted meat, Portuguese ruffians, and sea-sick Americans."[15]

Salted meat and hard bread formed the mainstays of the whaling diet, washed down with tea or coffee sweetened with molasses. By the end of the voyage, worms so infested the food that sailors dunked pieces of bread in their hot drink to "scald out" the vermin. A favoured dish was lobscouse, pieces of bread boiled in water with chunks of salt meat and seasoned with pepper, but the monotony of these staples defied the skills of even the most imaginative cook. Fresh foods supplemented the diet as they were available. After a visit to one of the Pacific harbours bags of oranges and bunches of bananas festooned a whaleship's rigging, while sacks of sweet potatoes, crates of squawking chickens and even the odd pig or goat foraging for scraps cluttered the decks. But these supplies ran out quickly on a long cruise and when Charles Wilkes visited the Pacific at the end of the 1830s he discovered that scurvy still flourished in the whaling fleet.

Appalling working conditions in the whale fishery were not completely the fault of greedy owners unwilling to share the profits of the trade. In a surprising number of cases, there simply were no profits to share. Enough vessels returned with rich cargoes to keep merchants interested in the enterprise, but it cost from $25,000 to $35,000 to send a whaleship to the Pacific and many returned with cargoes worth much less. According to Hohman, in 1850 one-tenth of all voyages were losing propositions.[16] And in some seasons it was much worse. In 1858, for example, a disastrous year, 64 percent of the voyages did not make their costs back.[17] No wonder so many merchants were abandoning the sperm whale fishery by this time.

For merchants, whaling was a gamble, but they at least had some expectation of making a profit. It is more difficult to understand why their crews hired on when prospects were so grim. One explanation is that they did not know what they were in for. Completely taken in by the false promises of the "landsharks," green hands set sail unprepared for the horrors of

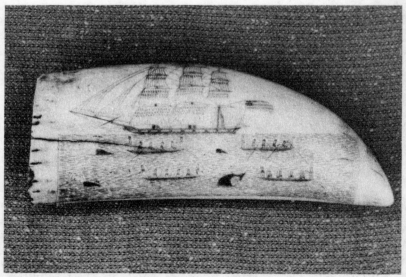

Scrimshaw. Whalers passed their leisure hours engraving on sperm-whale teeth or pieces of bone. This whaling scene decorates a large tooth.

the forecastle or the drudgery of their daily duties. If they returned safely, few could be convinced to repeat the voyage. Others sought to avoid some legal or family predicament. Still others were attracted by the promise of dangerous adventure or the lure of exotic places. ("I love to sail forbidden seas, and land on barbarous coasts," says Ishmael.) And then there were Indians, black Americans and destitute Portuguese islanders for whom any lay, no matter how small, exceeded what they could hope to earn at home.

Released from shipboard discipline and monotony when they arrived at the various Pacific ports, whaling men ran wild, with regrettable results for the native populations. Grog shops sprang up in all the ports to cater to the sailors' insatiable thirst for cheap liquor. Native women offered themselves to the sex-starved foreigners, either on the ships or, when these were declared off-limits, in rough brothels on shore. Venereal disease spread rapidly through the islands, along with the debilitating effects of alcohol. The natives also acquired guns from the newcomers, which they used in their tribal wars. Often whalers sailed away without paying for the supplies they had traded, leaving angry natives ready to wreak vengeance on the next foreign ship that happened to arrive. No wonder whaleships approached the islands cautiously with muskets at the ready. "There has been more done to destroy the friendly feelings of the inhabitants of the islands in the Indian and Pacific Oceans towards Americans, by the meanness and rascality of whaling captains," claimed J. Ross Browne after his cruise, "than all the missionaries and embassies from the United States can ever atone for."[18]

For many years the Bay of Islands in northern New Zealand was the most notorious harbour in the Pacific. Attracted by the presence of dozens of whaleships, an exotic collection of traders, sailors, deserters, escaped convicts and saloon keepers settled on the shores of the bay. By one count there were fifty grog shops in business by 1839, the same year that a visitor declared the settlement home to "a greater number of rogues than any other spot of equal size in the universe."[19] Missionaries complained of what they considered to be the corrupting influence of the seamen on the local Maoris. The whaling ships were brothels, they said, and the grogshops dens of depravity. Just as impor-tant, the outsiders sold guns to the Maoris, inflaming the warfare that periodically raged across the islands. Finally, in 1840, the British stepped in to impose their authority on New Zealand. The islands were annexed and suddenly the Bay of Islands was a foreign port for American whalers, with customs duties to pay and laws to obey. At the same time, the whaling fleet was extending its range in the North Pacific, making New Zealand a less convenient spot to refit.

While the Bay of Islands was the most popular whaling harbour in the southwest Pacific, the Hawaiian ports of Honolulu and Lahaina attracted ships cruising farther north. The mainland of Japan was closed to foreigners, so when whalers flocked to the Japan Grounds after their discovery in 1820 the closest place to refit was Hawaii. It became customary for ships to put in at the islands twice a year, once in March-April before the summer cruise to the north, and again in September-October before they headed south to cruise the southern hemisphere. At these times hundreds of vessels crowded the Hawaiian harbours, spilling thousands of sailors onto the waterfront where they traded for souvenirs, seduced the local women, drank and brawled. American missionaries, resident in Hawaii from 1820, despaired at the behaviour of their countrymen and in 1825 succeeded in convincing the local chiefs to forbid native women to visit the whaleships. In October, when the ships arrived, the sailors were outraged to learn of the new vice laws. At Lahaina on the island of Maui they blamed the missionary, Rev. William Richards, and threatened to kill him if the ban was not lifted. A mob marched on the missionary's home but the chief, Hoapili, posted a guard to protect his friend. There the incident ended. However, the following January the naval vessel USS *Dolphin* arrived in Honolulu harbour where its crew was less willing to put up with the new law. After negotiations failed to persuade Chief Kalanimoku, the commanding officer threatened to fire on the town and a gang of sailors assaulted Rev. Hiram Bingham. In the circumstances, the chief relented and allowed women to visit the ship. But the ban was reimposed soon after, and remained in place despite further instances of mob violence.[20]

The use of the Hawaiian Islands by whalers peaked after 1840 and continued to the Civil War period. Once the bottom dropped out of the trade for sandalwood, which flourished from 1815 to 1830, whaling became the cornerstone of the island economy. And along with their trade, the whalers continued to bring their seasons of violence. The worst example was the "great sailors' riot" of November 1852. A sailor named Burns, on shore leave from an American whaler, was arrested for drunkenness and tossed into jail at the fort, a large, wooden stockade on the outskirts of Honolulu. During the night Burns took a blow to the head in a melee and died. Two days later, after his funeral, a mob of sailors armed with clubs and fuelled with liquor ransacked the police station, set it ablaze, then roamed through the town on an all-night binge of looting and destruction. Finally a force of soldiers and militia, supported by the whaling captains in the harbour, cleared the streets and restored order.[21]

III

Seamen seeking permanent escape from the inedible meals, the substandard living quarters, the tyrannical officers, the boredom and the danger, faced two alternatives: either desert the ship or take it over. Certainly mutiny was the less common of these alternatives. The practical difficulties of organizing enough of the crew to overthrow the captain were immense, not to mention the question of where to escape the reach of the law. Once in a while, though, conditions on a whaler deteriorated enough to provoke rebellion.

In December 1822 the whaleship *Globe* departed Nantucket on a cruise to the newly-discovered Japan Grounds in the northwest Pacific.[22] Before sailing, Captain Thomas Worth formed a friendship with one of the ship's harpooners, Samuel Comstock. Comstock was a good-looking twenty-year-old from Nantucket who, since running away to sea seven years before, had shown an appetite for disreputable women and a dislike for discipline of any kind. On a previous voyage to the South Seas the young harpooner had been seduced by the charms of Polynesian life, and as the *Globe* made its way into the Pacific he carefully worked out a plan for leaving his old life behind him. Comstock was irritable, sometimes brutal, always unpredictable. Other members of the crew feared him, even as they respected his physical courage and his dexterity around a whaleboat. But Captain Worth for some reason liked Comstock and did not recognize the growing signs of trouble.

At the close of 1823 the *Globe* arrived in Honolulu after a disappointing cruise near Japan. Not only were whales scarce, but some of the provisions had gone bad and the crew had been living on short rations for months. To make matters worse, Comstock had been riding hard the men on his watch. Apparently he was goading them into jumping ship, and six of the crew obliged him during the stay in Honolulu. Since the *Globe* could not depart without a full crew, Captain Worth went ashore to find the deserters, taking Comstock with him. But the harpooner made sure that the missing men were not found, and eventually Worth decided to sign on replacements from among the foreign sailors hanging around the town. Comstock, one step ahead of his captain, had already plotted with four of these layabouts, who now joined the crew of the *Globe*.

Once again the ship set sail in search of whales, and once again tainted meat turned up among the stores. All the old grievances bubbled to the surface. The immediate cause of the mutiny was a thrashing Captain Worth gave to one of the new hands. Angered beyond endurance, a few of the sailors now wanted to desert in a boat. But this was the moment for which Comstock was waiting. Late on the night of January 26, accompanied by three other members of the crew, he quietly made his way back to the captain's cabin. Seizing an axe, Comstock stood over the sleeping figure of his trusting friend and split his skull open with a single blow. Then the chief mutineer fell on the three mates, almost decapitating one with the axe, stabbing another with a bayonet and executing the third with a gunshot to the head. Soaked in blood and spattered with gore, a wild-eyed Comstock emerged from the officers' cabins in command of the ship. He and his cohorts had all the weapons; the rest of the crew could not have resisted had they wanted to. After impaling the corpse of Captain Worth on a long sword, the mutineers tossed the four mutilated bodies into the sea, and Comstock set a course for the west where he planned to settle on a lonely island, then sink the ship. Since he was the only person on board who knew how to navigate, the others had little choice but to defer to his command.

In mid-February the *Globe* arrived at the island of Mili in the Marshall Group. While the crew unloaded the stores, Comstock visited a native village near the anchorage. Unknown to the others, he was now putting into effect stage two of his plan. By winning over the natives with generous gifts, Comstock hoped to convince them to murder all the other seamen. But this time his planning failed; the suspicious sailors shot their crazed leader before he could win the allegiance of the natives.

The night of the murder, natives and sailors indulged in a raucous party. But beneath the festivities, the plotting continued. Led by harpooner Gilbert Smith, six of the crew returned to the ship and under cover of darkness sailed away from the island. After a four-month crossing of the Pacific, they brought the *Globe* into Valparaiso harbour where they were immediately clapped in jail. After close questioning by the American consul in Valparaiso, the six men sailed home to Nantucket where all but one were exonerated. The sixth man, Joseph Thomas, was held for trial but eventually was acquitted. Meanwhile, the whale merchants of New England urged the federal government to send the navy after the mutineers, to punish the guilty and rescue any innocent survivors. They did not know, of course, that shortly after the departure of the *Globe* from Mili the natives killed the remaining crewmen in revenge for the mistreatment of two island women. Two only were spared, and were adopted into the homes of the villagers.

When at last the United States navy came looking for them almost a year and a half later, they found the only two survivors of the *Globe* mutiny dressed in loincloths, their skin burned black by the sun, their hair dangling in long ringlets, indistinguishable from the natives who seemed to treat them as family. When the story of the fate of the other crewmen came out, the naval commander pardoned the residents of the island for their part in the affair, but warned them that if ever they murdered any visiting white men again a ship would return and destroy the island and everyone on it. That settled, the two young survivors returned to New England aboard the naval vessel to write a book about the mutiny.

Mutiny provided an extreme solution to the problems facing disgruntled seamen. Desertion was a simpler alternative. As crews on whaling ships became more inexperienced, less able to endure the many months at sea, the incidence of desertion increased. Few vessels returned home with the same crew they had when they set out; by the 1840s and 1850s almost two-thirds of an average whaling crew abandoned ship during a voyage.[23] Many of these men were escaping from a brutal captain, or simply from the monotony, discomfort and danger of the whaler's life. Others fell under the spell of a potent myth, that the tropical islands of the South Seas offered an earthly paradise, a chance to start a new life, to try on a new identity. "Indeed, there does not occur a greater difficulty to European ships in the South Seas, than that of keeping their crew together," remarked one experienced sailor, "such is the seduction of that life of indolence and carelessness which the several islands hold out. The beauty of the country . . . and still more, the facility with which the necessaries of life may be procured, are, in general, temptations too powerful to sailors exhausted with the fatigue of so long a voyage: add to this the women; then the difficulty of retaining our seamen against so many attractions will excite no further surprise."[24]

Deserters from whaleships joined a growing community of beachcombers–convicts on the fly from the settlements of Australia, runaway seamen from the trading vessels, even the odd missionary gone native–scratching a living among the Pacific islands. "They are, mostly, a reckless, rollicking set," wrote Herman Melville in his novel *Omoo*, "wedded to the Pacific, and never dreaming of ever doubling Cape Horn again on a homebound passage. Hence their reputation is a bad one." Melville was describing the hard-core beachcomber, usually an escaped convict, who had no future back in England or the United States. The majority were not like this at all. They were deserters who learned that life in paradise could be difficult and boring. Few outsiders established themselves in indolent ease on the islands. More usually, they ran afoul of the local natives or grew disenchanted with an aimless life apart from family and friends. "I grew weary of lying all day long in the shade, or lounging on the mats of the great house, or bathing in the bright waters," wrote one disillusioned runaway.[25] Usually they cut short their stay after a few months and signed on with a passing ship. From this pool of beachcombers the whalers hired replacements for the crewmen who regularly deserted.

As Melville noted, beachcombers enjoyed a nasty reputation among respectable missionaries, merchants and whaling captains. In 1825, for example, the whaling masters of Nantucket petitioned their government to do something about American deserters living in Hawaii. There were more than 150 of them, the captains reported, "prowling about the country, naked and destitute, associating themselves with the natives, assuming their habits and acquiring their vices." If the U.S. navy did not go in and clear them out, Honolulu would "soon become a nest of pirates and murderers." In response the USS *Peacock* arrived in Honolulu in October 1826, and after urging the local chiefs not to harbour deserters, Captain Thomas Jones removed thirty beachcombing Yankees from the island.[26]

But beachcombers were not as negative an influence in the Pacific as some observers made out. In many instances they played an important role as cultural middlemen, teaching useful skills with the new metal tools, acting as interpreters and business agents, and generally explaining the ways of the white intruders to the curious natives. In the early decades of the nineteenth century, until the permanent settlement of traders and missionaries, beachcombers were the most influential white presence in the islands of the South Pacific.

The positive role of the beachcomber can be extended to the impact of sperm whalers on the South Pacific generally. Until recently, most writers have agreed that the whalers' influence on native occupants was almost totally negative. The title of Alan Moorehead's 1966 book on the South Pacific, *The Fatal Impact*, came to summarize an entire approach to the issue of cultural contact. According to the supporters of the "fatal impact" theory, whalers, along with missionaries, explorers and traders, spread disease, liquor, guns and alien religions through the region, depleting island populations and corrupting innocent lifestyles. Primitive native cultures were apparently no match for the overwhelming superiority of Western technology and the attraction of its vices. More recently, however, scholars have debunked aspects of the "fatal impact." There is no evidence, for example, that the population of the South Pacific declined dramatically after its contact with outsiders, or that a large number of natives became demoralized layabouts from consorting with American sailors. Instead, scholars emphasize the creative responses made by native culture to the new opportunities presented by the arrival of whalers and others. Metal tools were adapted to island agriculture; whaleboats proved superior to canoes for certain purposes; natives travelled widely to other parts of the Pacific and returned with new knowledge and experience; and a healthy demand for goods and services by the whalers created new economic opportunities. Other results of contact are too numerous to mention. Naturally they were not all positive. But neither, from the natives' perspective, were they all negative. Rather, the whaling era was a period of mutual exploitation when whaler and Pacific islander each tried to take advantage of opportunities presented by the other. Such encounters are inevitably filled with turmoil, but they are not necessarily destructive and are just as likely to be invigorating.

IV

Sperm whaling reached its peak of productivity in the early 1840s when an average of about 160,000 barrels of sperm oil arrived in New England each year. Then it began a steady decline, as year by year sperm whales accounted for a smaller percentage of the total American catch and captains took their ships after other species. Scientists still disagree about the cause of this decline. Had whalers so depleted the stocks of sperm whales in the Pacific and Indian oceans that hunting them was uneconomical? Certainly some observers believed so. Forced to finance longer and longer voyages to fill their holds, whale merchants grew dissatisfied with their meagre profits and withdrew their capital in order to invest in other sectors of the economy. Other observers felt that sperm whales had not so much declined dramatically in numbers as they had become skittish and difficult to capture. Either way, the result was the same.

But there were other factors. In 1835 a French whaleship discovered an area teeming with right whales off the coast of what is now British Columbia and within a few years more than a hundred American ships were spending their summers on the Northwest Grounds north of the fiftieth parallel of latitude.[27] Right whales were desirable because they produced more oil than sperm whales and their jaws held rows of baleen. Then, in 1848, Captain Thomas Roys sailed his bark *Superior* through Bering Strait and opened a rich bowhead whaling ground to the American fleet. The bowhead had the largest mouthful of baleen of any of the whale species and discovery of its summer grounds came at a particularly fortuitous moment. From about 1830 a series of substitute products began to challenge the popularity of whale oil. As illuminants, camphene, derived from turpentine, and coal gas made their appearance in homes and factories. As lubricants and in industry, fish and seed oils, especially cottonseed and rapeseed, proved increasingly effective. Finally, the discovery of petroleum in 1859 dealt a major blow to the whalers and led to a drastic decline in the demand for animal oil.

However, at the same time as whale oil was losing its place in the market, the value of baleen was on the rise, principally because of changes in women's fashions that required large amounts of the flexible substance as reinforcement in hoop skirts and undergarments. Between 1840 and 1844 the price of baleen doubled, and it remained strong thereafter. The sperm whale, once the most valuable animal in the sea, carried none of this valuable baleen in its mouth and produced much less oil than the bowhead or right whale. It now took second place to these other species inhabiting the waters around the rim of the North Pacific, and in the middle of the nineteenth century the focus of the whaling industry shifted inexorably northward toward the ice-bound reaches of the Arctic.

EXTERMINATING THE DEVILFISH

The mammoth bones of the California Gray lie bleaching on the shores of those silvery waters, and are scattered along the broken coasts, from Siberia to the Gulf of California; and ere long it may be questioned whether this mammal will not be numbered among the extinct species of the Pacific.[1]

–Captain Charles Scammon, 1874

During the heyday of Pacific sperm whaling one of the several grounds frequented by the whaleships was off the coast of Baja California. After leaving Hawaii each fall on their southern voyages, many whalers chose to head for the Baja where they pursued sperm and humpback whales during a leisurely cruise south toward the equator. As British and American whalers skirted the long peninsula, putting in occasionally to take on water and meat at the remote Spanish ranches along the low, arid coastline, they saw pods of another species of whale that as yet did not arouse their killer instinct. The Pacific gray whale arrives at the warm-water lagoons of Baja and the adjacent Mexican shore every winter as regular as clockwork. While the females enter the lagoons to give birth and nurse their young, males and youngsters loll about the entrances, courting and feeding and waiting for the beginning of the long migration north in the early spring. These slow, methodical animals presented a tempting target for the whaleships, but for many years they did not repay the effort it took to catch them. The blubber of the gray whale produced an oil inferior both in quality and quantity to that of other species, and the animal's

113

whalebone, measuring only about eighteen inches in length, grew pitifully small compared to the much larger pieces in the mouth of the right and bowhead whales. As long as it made no economic sense to pursue the gray whale, whalers left this great mottled beast in relative peace.

The California gray whale is the last survivor of a species that formerly roamed across much of the northern hemisphere. In the Atlantic it was called the scrag whale by early New England settlers, a name that referred to the line of knobby bumps "scragged" along the ridge of the animal's back in place of a dorsal fin. But this stock no longer exists, apparently disappearing near the close of the seventeenth century.[2] Another herd of grays, inhabiting the northwest Pacific between Korea and the Sea of Okhotsk, is on the verge of extinction. Severely depleted by Japanese and Korean whalers in the early to mid-twentieth century, this herd now numbers a few hundred, at most.

The last healthy herd of grays occupies the waters off the west coast of North America. These animals are remarkable for the long migration they undertake every year between their summer feeding grounds in the Bering and Chukchi seas and the calving lagoons of the Baja Peninsula, about six thousand miles to the south. The longest annual migration by any mammal on earth, it begins in October when the northern ice crowds down from the Arctic. For the previous five months the animals have been gorging themselves on small organisms living on the sea floor. It has been estimated that a gray whale devours more than a ton of these tiny crustaceans every day. A feeding whale glides on its side across the sea bottom, sucking up vast mouthfuls of sediment that is strained through the baleen. It swallows the nourishing food and expels the debris in long plumes of mud that are seen trailing in the water. On their journey south the whales do not stray far from shallow coastal waters. Travelling in small groups, they make about seventy miles a day, pausing occasionally to feed and to mate, until late in December they begin arriving at the nursery lagoons in Baja.

The whales were not allowed to make their coastal journey entirely unmolested. Aside from their one natural enemy, the killer whale, the grays had been pursued by aboriginal hunters since before recorded history began. Embarking in their skin and wood boats, natives fearlessly attacked the giant animals with stone-tipped lances and harpoons made of mussel shell. In

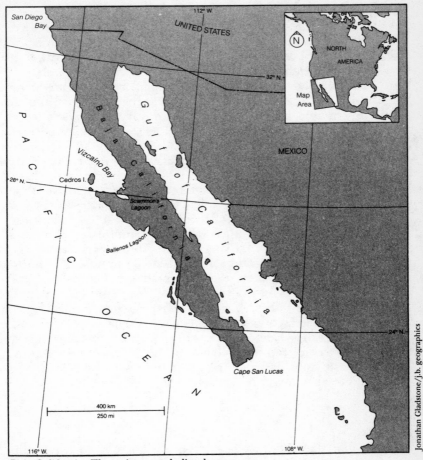

Baja California. The main gray whaling lagoons.

Alaska the Aleuts and Eskimos may have dipped their lances in poisonous aconite made from the monkshood plant. Once they had stabbed the whale, the hunters withdrew to shore to wait for the poison to take effect and the corpse to float in on the tide. Farther south, the Nootka Indians on the west coast of Vancouver Island and the Indians on Washington's Olympic Peninsula darted harpoons at their prey. A long line of cedar bark and sinew trailed behind the harpoon, dragging several sealskin buoys to tire the fleeing animal. Once the kill was made, the carcass had to be towed back to the village, a task that often took

115

several days if the wounded whale had headed out to sea.

Whaling played an important role in aboriginal culture, both for the use the people made of whale products and the prestige the hunt conferred on the male members of the tribe. It was cloaked in elaborate ritual and festivity. But the number of gray whales killed by natives was small and certainly posed no threat to the existence of the species. This threat did not appear until the middle of the nineteenth century when, suddenly, everything changed for the gray whale. As sperm whales became harder to find, whalers decided it no longer made sense to ignore the plentiful grays whose regular habits made them easy to hunt. The pioneers of gray whaling were a pair of ships from Connecticut that entered Magdalena Bay on the west coast of the Baja at the end of 1845 and killed thirty-two whales. That was just the beginning. In an orgy of destruction lasting three decades the population of grays was nearly annihilated. All along the sunburned coast of the peninsula hunters invaded the shallow calving lagoons and slaughtered cows and calves indiscriminately. Because such a disproportionate number of the victims were females and calves, the total population of the herd plummeted much faster than other whale species. In 1874, not thirty years after lagoon whaling began in Baja, Charles Scammon, a veteran whale hunter whose report on the fishery remains a classic, reported that "the large bays and lagoons, where these animals once congregated, brought forth and nurtured their young, are already nearly deserted."

I

It is ironic that Charles Scammon should have written this epitaph for the gray whale, since he played such an important role in its near-extermination. Originally from Maine, Scammon made a migration of his own late in 1849 when he sailed in command of a New England merchantman around Cape Horn to San Francisco. Adopting the booming gold rush centre as his new home, Scammon began sailing out regularly on whaling and sealing excursions and trading voyages to China and South America. His attention was attracted to the gray whales that he saw gathering along the Baja shore late each autumn. He knew that the whales frequented Magdalena Bay, and he joined the

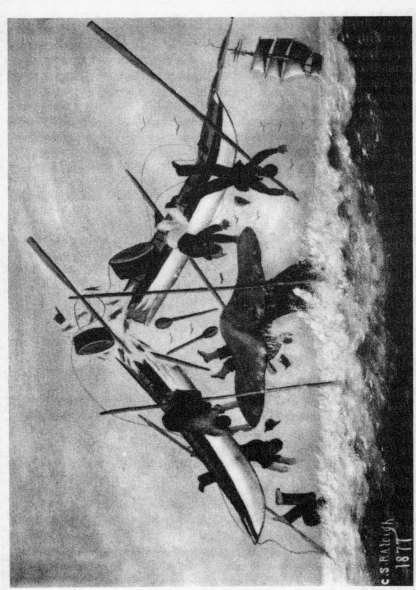

With a flick of its giant tail, a whale smashes a boat. A painting by Charles S. Raleigh.

hunt there in the mid-1850s, but it was not until 1857 that Scammon stumbled on the fateful discovery that made him famous.

"During the spring and summer months of 1857," he later recalled, "we were engaged with the brig *Boston* in whaling, sealing and Sea Elephant hunting, but with ill success." Facing an unprofitable season, Scammon decided to investigate stories he had heard that a large number of whales could be found in a previously unexploited lagoon at the bottom of Vizcaino Bay, a large body of water about halfway up the west coast of the Baja. The lagoon was no secret to seal hunters who had been active in the area for decades, but whaling captains had been discouraged by the shallow water, the treacherous currents and the long sandbar that fronted the lagoon and reduced its entrance to a narrow passage. With nothing to lose and a willing crew, Scammon decided to take a chance. He sent to San Francisco for a small schooner to join his 181-ton brig, and both vessels set out at the end of November in search of the new whaling ground.

The lagoon toward which Scammon sailed is located on a low, desolate stretch of Baja coastline. It was known to Indians who had been fishing there since prehistoric times, and to Mexicans chiefly for a fresh-water spring located at one end, but it was isolated by the Vizcaino Desert from the trickle of traffic up and down the Baja peninsula. Between the lagoon and the desert were salt beds formed over time by the regular precipitation of salt from the shallow pools of water left by the falling tide. Later, Mexicans would arrive to mine the salt and ship it for sale to San Francisco, but in 1857 it was whales, not salt, that attracted the attention of the outside world.

When Scammon arrived in the vicinity of the lagoon, he sent the schooner and three whaleboats to locate the entrance and determine if it was deep enough to allow the *Boston* to get through. Two days later a boat returned with the good news that the schooner was already in the lagoon and that there was plenty of water for the brig. The next day the *Boston* made its way through the passage and eventually to the head of the lagoon where Scammon found what he was looking for—large numbers of female gray whales swarming in the shallow waters, preparing to give birth and nurse their newborns. The captain had discovered the most important of the Baja nurseries. In the years ahead his fellow whalers would commemorate this pioneering voyage by naming the lagoon after him. It was a dubious honour.

In the seven seasons that followed Scammon's arrival, about six hundred gray whales were killed in his lagoon, along with many more newborns sacrificed as lures to capture their mothers. This number is less than 10 percent of the total pre-1875 gray whale slaughter along the California and Mexican coasts, but Magdalena Bay was the only other single location where so many whales were killed in such a short time. The lagoon that Charles Scammon saw that January morning in 1858, crowded with the backs of gray whales jostling in the narrow channels, was, within a decade, virtually empty of the animals, and the name of Scammon's Lagoon has survived not as a tribute but rather as a reminder of one of the bloodiest episodes in the history of the whale hunt.

For all its later success, the hunt in Scammon's Lagoon began with difficulty. The whales were confined in a restricted area, the water was placid and shallow; the hunt seemed all too easy. But, as Scammon's men soon discovered, the channels of the lagoon were swept by perilous currents and eddies that made manoeuvring their boats almost impossible. At high water, boats strayed into the shallows only to be stranded as the tide suddenly withdrew. The movement of the whales stirred up the silty bottom so that the whalers could not keep track of their prey. ("I didn't know, sir, that the whale was within fifty fathoms of me," complained one harpooner to his captain, "when up we went—and there ain't enough left of the boat to kindle the cook's fire.") And most unnerving of all, the female gray whale turned out to be a very dangerous animal when angry or fearful for its young, smashing the wooden boats with its tail flukes and dumping the whalers into the warm water, leaving them with bruised courage and broken limbs. "I have a strong notion they are a cross 'tween a sea-serpent and an alligator," concluded one harpooner looking over his staved boat. Of all the names whalers applied to the California gray whale—"Mussel-digger", "Grayback", "Hard Head" and "Rip-sack" among them—it was as the Devilfish that it was most commonly known, and it came by the nickname honestly.

On the second day of whaling all these difficulties became clear to Captain Scammon and his men. The hunt had hardly begun when two of their three boats were scuttled by whales, leaving two men with fractured limbs and nearly half the crew injured and unable to continue. Those who were not actually hurt were so frightened at the prospect of returning to the hunt that whaling had to be postponed for several days. Then, thinking his men had overcome their fear, the captain dispatched another boat, only to watch in dismay as the oarsmen all jumped overboard when they closed with a whale.

Faced with the prospect of having to abandon the hunt, Scammon hit upon a technique that subsequently became the hallmark of lagoon whaling. This technique depended on the use of the bomb-lance, a deadly weapon recently introduced to the whalers' arsenal. It was an explosive missile, about twenty-one inches long, armed with a time-delay fuse and fired from a squat, large-bore shoulder gun by the mate standing in the bow of the whaleboat. Detonating deep in the animal's interior, it caused a terrible wound and sometimes instant death. At sea the bomb-lance was used to dispatch a whale already harpooned and exhausted from the chase. It allowed the final kill to be made from a safe distance instead of up close with a hand lance where the animal might smash a boat in the agony of its death flurry. Scammon, however, adapted the weapon for his own purposes. He placed his boats near where the whales would be passing, but in shallow water where the vengeful animals could not reach them. The idea was to "bomb" the whales once, twice, even three times until they died from massive internal injuries. And it worked. That first day, marksmen planted bombs in three whales. One died instantly; lookouts spotted the corpses of the other two the next morning floating not far away.

"From that time, whaling was prosecuted without serious interruption," Scammon reported. "The try-works were incessantly kept going . . . until the last cask was filled. Nor did we stop then; for one side of the after-cabin was turned into a bread-locker, and the empty bread-casks filled with oil; and the mincing-tubs were fitted with heads, and filled, as well as the coolers and deck-pots; and, last of all, the try-pots were cooled, and filled as full of oil as it was thought they could hold without slopping over in a rough sea." In all, Scammon packed away seven hundred barrels of oil before the whales left the lagoon that spring to start their migration north to the Bering Sea. It was a rich cargo that he took home to San Francisco, but a trifling one compared to the hundreds of thousands of barrels that would be produced by the slaughter of the gray whales in the years ahead.

II

Before the year was out, Scammon returned to his lagoon, this time in command of "a little squadron of vessels," as he put it, the three-hundred-ton bark *Ocean Bird* and two small schooner tenders. Word had spread through the fleet, and he was not alone. "Although this newly discovered whaling-ground was difficult of approach," he wrote later, "and but very little known abroad—and especially the channel which led to it—yet, soon after our arrival, a large fleet of ships hovered for weeks off the entrance, or along the adjacent coast. . . ." Of these, six joined the *Ocean Bird* inside the lagoon for a season of whaling. By the spring of 1859, when the whales again departed, the whalers had taken two hundred animals, ten times as many as Scammon accounted for in his first season at the lagoon.

The presence of seven whaleships meant that twenty-five to thirty boats were cruising the lagoon daily. Some followed Scammon's example and "bombed" their prey but most employed more traditional methods, pursuing the whale into shallow water, making fast with the barbed harpoon, then hanging on tightly while the stricken animal fought desperately to escape. Several of the boats might be fast to whales at the same time, their lines crossing and re-crossing, their mates hurling abuse and instructions as they manoeuvred frantically to keep from colliding with each other, at the same time trying to avoid the flukes and jaws of their prey. This "scene of slaughter" was, Scammon thought, "exceedingly picturesque and unusually exciting," and was made even more so when the light on a calm morning played tricks and threw up distorted images of the chase. "At one time, the upper sections of the boats, with their crews, would be seen gliding over the molten-looking surface of the water, with a portion of the colossal form of the whale appearing for an instant, like a spectre, in the advance; or both boats and whales would assume ever-changing forms, while the report of the bomb-guns would sound like the sudden discharge of musketry. . . ."

The hunt generally began early in the day and by noon the boats were hauling their catch back to the ships for flensing. Once stripped of their bone and blubber, the carcasses were cut loose to float wherever the eccentric currents carried them, food for the sharks roaming lazily through the lagoon. Those that did not drift out to sea grounded in the shallows where the falling tide stranded them, grotesque, ragged mounds of bone and tattered flesh. These abandoned carcasses, called "stinkers," attracted the attention of Mexican traders who were gathering at the whaling harbours. The organs and bones of the dead animals still contained some oil, which the scavengers obtained by boiling them in trypots on shore. This practice, known as "carcassing," supported tiny, nomadic communities of Mexicans who sold the oil back to the ships along with food supplies and locally-produced mescal liquor.

Scammon's Lagoon remained an important whaling centre for only two more seasons. By then the population of grays was so depleted that most vessels went elsewhere. Captain Scammon, for example, cruised farther south along the Baja coast to pioneer another rich hunting ground at San Ignacio Lagoon. But everywhere it was the same story; by the early 1860s the whales were disappearing as the lagoon-based phase of the gray whale hunt came to a close. In Scammon's case, 1863 was his last season as a whaler. For the next decade he served in the U.S. marine revenue service, then retired to a farm in California in 1874, the same year that he published his book, *Marine Mammals of the Northwestern Coast of North America*. He lived until 1911, but by then the whaling industry in the lagoons of Baja must have been only a dim memory.

III

Out along the coast in the migration routes and off the mouths of the lagoons where the male grays congregated, there were still enough whales to attract a sizeable fleet of whaleships until the late 1860s. And even as these ships began to drift away from the gray whale grounds, there remained one more group of hunters to carry on the hunt. These were coastal whalers, based at shore stations, who came out into the migration lanes in boats to prey on the animals as they passed on their annual journey. Shore-based whaling began at Monterey in 1854 and spread to harbours along the coast of California, then south to Baja. The boats, which roamed up to ten miles from shore, were fitted with a Greener's harpoon-gun mounted on a swivel in the bow and firing a harpoon four and a half feet in length. According to Scammon, these weapons were accurate up to thirty yards and were the only thing that made coastal whaling possible, since the whales had become "exceedingly wild and difficult to approach" in the usual manner with the hand harpoon. Boats towed the dead whales back to the shore station where resident crews, mainly Portuguese immigrants from the Azores and Cape Verde Islands, stripped the blubber and boiled the oil. As well as grays, these stations took humpbacks, the famous "singing" whales, known for their underwater vocalizations and long flippers. But even so, by the 1880s whalers were feeling the disappearance of the grays and stations were closing all along the coast. The last one barely survived into the twentieth century.

David Henderson, the main historian of this fishery, has calculated that by the mid-1870s only about two thousand California gray whales remained alive, from a pre-whaling herd of fifteen to twenty thousand animals. But the scarcity of whales was not the only reason for abandoning the hunt; there were compelling economic reasons as well. Petroleum had been replacing whale oil as an illuminant for several years and as the price of oil dropped, the price of baleen was steadily rising. As oil became less profitable, it made more sense for whalers to turn their attention to animals that gave a better quality baleen than the gray.

With only two thousand of the animals left alive, it is not surprising that for many years people believed the gray whale to be virtually exterminated. Hunters no longer came to the calving lagoons of Baja and the surviving animals apparently kept far enough out to sea to avoid the notice of coastal marine traffic. Deep-sea whalers operating from modern factory ships took about a thousand grays in the years after 1914, so their future was still in doubt in 1937 when the gray whale was protected from hunting by international agreement. By 1951 all the whaling nations had added their signatures to this agreement and the gray was fully protected from commercial whaling. However, many people felt that the ban had come too late, that in commercial terms, at least, the species was already extinct.

It turned out that reports of the gray whale's demise were premature. In 1946 Carl Hubbs at the Scripps Institution of Oceanography at La Jolla, California, stationed his students on the rooftops of buildings on campus to count the whales as they migrated down the coast. By 1952 there was an official government headcount underway. What these early census takers proved was that the gray whale had not been wiped out—that in fact, the population was growing. Today that population numbers about twenty thousand animals. In other words, about as many California gray whales exist today as lived 150 years ago, before the commercial hunt began.

The gray whale, so close to disappearing not long ago, is now one of the most commonly seen whales in the world thanks to the phenomenal growth of whale watching along the Pacific Coast. Because it sticks so close to land, more than two million people view the herd every year. And, ironically, one of the most popular places to see these giants is none other than the lagoons of Baja. In January 1972 the Mexican government declared Scammon's Lagoon (Laguna Ojo de Liebre) the world's first whale refuge and followed up by placing a similar status on Laguna San Ignacio and Magdalena Bay. Access to the lagoons by motorboat is tightly controlled but each year growing crowds of sightseers arrive to get a look at the whales. From slaughterhouse to sanctuary, the Baja lagoons have come full circle.

Chapter Seven

REGIONS OF ETERNAL FROST

We seem to be dwelling in some haunted house filled
with unearthly and mysterious noises. We sit like hares,
startled and alarmed at the slightest sound, dreading
and fearing we know not what.[1]

> —Charles Edward Smith, surgeon aboard
> the ice-bound whaler *Diana*, 1866

Though the British gradually abandoned sperm whaling in the
South Pacific during the 1820s, they did not give up whaling
altogether. Even as their southern fleet dwindled in size, the
fleet of British whaleships sailing to the Arctic made an important discovery in Davis Strait that enabled that branch of the
industry to embark on a period of great prosperity—at the same
time bringing the population of bowhead whales in the eastern
Arctic to the brink of extinction.

From the beginning of whaling in the Davis Strait in the
early 1700s, ships were confined to a narrow band of open water
along the west coast of Greenland. The way across to the coast of
Baffin Island was blocked by a dense mass of ice, called the
"middle pack," which collected in Baffin Bay and Davis Strait. So
thick was the ice, and so variable the winds and weather, that for
more than a century whalers dared not risk their wooden ships
by venturing into it. In 1817, however, two British ships, the
Larkins of Leith and the *Elizabeth* of Hull, forced a passage
westward across the top of Baffin Bay at latitude 77 degrees
north, hundreds of miles north of the Arctic Circle. The two
captains emerged from the pack into an area of open water

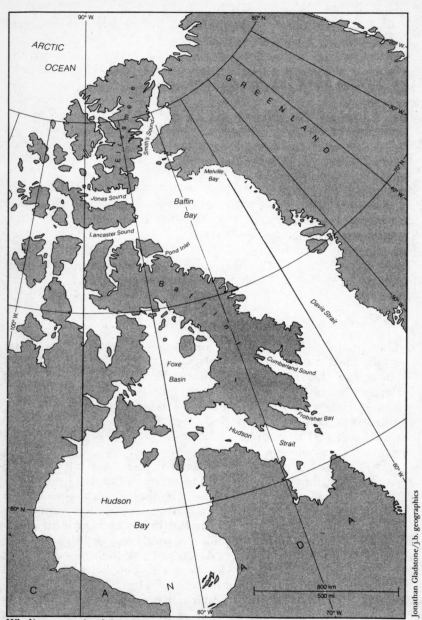

Whaling grounds of the eastern Arctic.

Jonathan Gladstone/j.b. geographics

called the "west water" that, ironically, cleared of ice earlier in the season than more southerly latitudes. The "west water" turned out to be a rich new hunting ground where previously the whales had been able to escape the harpoons of their hunters.

This was the first recorded crossing of Baffin Bay by any whaler. The next year, 1818, a British naval expedition appeared on the whaling ground, the first in a long series of British attempts to find a Northwest Passage through the Arctic archipelago. Led by Commander John Ross and consisting of two sailing vessels, the expedition crossed the top of Baffin Bay, then ventured a good distance into the broad mouth of Lancaster Sound before exiting the Arctic down the east coast of Baffin Island. These pioneering voyages by whalers and explorers proved that the "middle pack" was not impassable, and that once they committed themselves to the "west water" ships could escape home again before the onset of winter. Whalers immediately adopted this circular route into and away from the whaling grounds on the west side of Baffin Bay.

As a result, the number of animals killed rose dramatically. But the price the whalers paid for this bounty was high. By extending their voyages deeper into the Arctic, they increased the risk of losing their ships in the ice. As the catch increased, so did the number of vessels sent to the bottom and the number of sailors dead from disease and exposure.

I

Nowhere were as many whaleships destroyed as in Melville Bay, a broad sweep of water on the Greenland coast at the top of Baffin Bay. Because the bay is shallow and edged with many rocky islands, ice gathers thickly there and holds fast to the bottom. In calm weather it is a white, featureless plain, its flatness broken only by the occasional towering iceberg or, toward land, by clumps of rock that emerge through the surface, as one sailor remarked, "like the uplifted hands of drowning men."[2] On a clear summer day, despite a temperature well below freezing, the bright sunshine reflected off the ice and water so strongly that it scorched the exposed faces of the whalemen as they went about their duties.

Between this shorefast ice and the dense, shifting "middle pack" to the south, a passage sometimes opens. Across this ribbon of water the whalers made their way westward. It was by

Working through the ice. Arctic whaling crews had to saw a passage through the ice pack using giant saws mounted in triangular frames. Other crewmen laboriously turned the capstan to haul in ropes attached to hooks planted in the ice.

no means a safe passage. The fate of the whaleships rested with a capricious wind. When it blew off the land from the north, it moved the floes away from the edge of the shorefast ice, making thin openings of clear water for ships to squeeze through. On the other hand, Robert Goodsir, a surgeon who made the crossing in 1849, described what happened when the wind shifted. "Should a south-west or southerly wind set in whilst they are slowly working their way through, between the land-ice and the loose floes, it frequently drives in the middle ice upon them, with such violence and rapidity, that the vessels are crushed between them like egg-shells." When the two packs collided, Goodsir wrote, "it was as if one was standing over the site of an earthquake. The ponderous ice, trembling and slowly rising, would rend and rift with a sullen roar, and huge masses, hundreds of tons in weight, would be heaved up, one above the other, until, where it was before a level, an immense rampart of angular blocks became piled."[3] Dozens of ships were destroyed in the ice of Melville Bay. Whalers called it the "Breaking-Up Yard," and as they worked their way across they kept a bundle of clothing and possessions close by so they could abandon their vessel at a moment's notice.

Even when weather conditions were favourable, sailing ships made slow progress across the bay. If leads were wide enough, crews launched the whaleboats and took their vessels under tow. When the cracks narrowed to barely a ship's width, the men strung themselves out on the ice and, hauling on ropes, trekked slowly westward. "The ships used to follow each other closely with jib-boom projecting over the taffrail of the preceding ship," recalled a veteran of the early days, "the crews tracking along in full chorus or to the enlivening notes of fiddles, bagpipes, fifes and drums. Thus you might see fifteen or twenty ships following each other in a string."[4] When a storm blew up, or the two continents of ice closed together, the ships took refuge in docks hastily sawed in the floes.

The Breaking-Up Yard was the scene of many disasters, but the worst occurred in the summer of 1830. A total of nineteen ships went down, one-fifth of the entire British fleet in Davis Strait that season. Most sank in Melville Bay attempting to get across to the west side.

On June 10 near the Devil's Thumb, the tall column of rock on the Greenland shore that roughly marked the beginning of the treacherous crossing, about fifty whaleships were patiently waiting for a change in the weather to allow them to proceed farther. Suddenly a crack opened in the pack and twenty-three vessels were able to hurry into the ice before the lead closed behind them. As they struggled ahead the squadron broke up into several smaller groups that were swept away in different directions. On June 19 one group of six vessels became mired in the ice in a single column south of Cape York and could advance no farther.

The evening of June 24 blew a fierce gale down onto the ships. The sky darkened. Sheets of hail and driven snow battered the imprisoned whalers. Driven by the wind, the ice began to buckle and heave and close in tightly around the vessels. One large floe came rushing down upon the first ship in the column, the *Eliza Swan*, and struck it a terrific blow in the bow. But instead of crushing the hull, the ice lifted the ship and hurled it back against its neighbour, the *St Andrew*. Then it passed completely beneath the *Eliza Swan* and slid along the length of the *St Andrew*, staving in about twenty planks but leaving the vessel afloat.

The other four ships enjoyed no such luck. The ice overwhelmed them and in just fifteen minutes they were reduced to fragments. "The scene was awful," wrote Sir John Leslie, who had his account from eye-witnesses, "the grinding noise of the ice tearing open their sides; the masts breaking off and falling in every direction; were added to the cries of two hundred sailors leaping upon the frozen surface, with only such portions of their wardrobes as they could snatch in a single instant."[5]

Elsewhere in the ice the storm claimed seven other ships. Then, early in July, another gale blew up and destroyed five members of the original fleet that had been left behind at Devil's Thumb. A surgeon on one of these stricken vessels later reported that he saw two converging ice floes crash through opposite walls of his cabin to meet in the middle of the room.

When the weather cleared, more than a thousand sailors were living on the ice in Melville Bay, refugees from the ships that had gone down or were slowly sinking. Living in tents or underneath whaleboats, these men began a carouse that lasted three weeks and became legendary as the "Baffin Fair." Enough supplies were retrieved from the ships that no one was in danger of starving, and enough ships remained afloat to carry everyone home when the ice relaxed its grip. Meanwhile, the liquor casks were opened and the men, liberated from shipboard discipline, enjoyed a holiday. For all the destruction, the only lives that were lost that season were those of drunken sailors who wandered away from the tent cities and died of exposure on the ice.

Arctic whalers customarily burned any ship that foundered in the ice, in order to get at the valuable supplies in the hold and to remove the hazard to navigation. At night during the "Baffin Fair" red flames shot skyward from the blazing hulks, outlining the dark shadows of carousing sailors and lending a diabolic air to the carnival.

When the ice loosened, the marooned sailors dismantled their camps and found berths on the ships that remained seaworthy. The pack remained almost impenetrable and it was late August before the first ships reached open water on the west side. By that time the whaling season was almost over and it was time to head for home. As a result, in addition to the toll of nineteen vessels destroyed and twelve crippled, twenty-one ships returned to Britain in the fall of 1830 with no cargo.[6]

II

The British whaling industry recovered from the losses of 1830. For the next few years returns were higher than ever. But then in 1835, and again in 1836, the industry suffered a pair of disastrous seasons that marked a turning point in the Davis Strait fishery.

The Breaking-Up Yard was not the only obstacle to a successful voyage. After chasing whales near the entrance to Lancaster Sound and off Pond Inlet in July and August, the fleet customarily straggled down the coast of Baffin Island and out of Davis Strait in September. When conditions were right, the ice by this time had moved away from the shore, opening a passable channel south. But conditions were not always right. When the wind blew from the east and the northeast the ice piled in against the rocky coast, catching the whaleships before they could get away.

In the summer of 1835 the fleet encountered extremely heavy ice in Baffin Bay. As the days passed and the pack in Melville Bay held firm, impatient whalers forced a passage through the dense "middle pack" farther south. Arriving off the coast of Baffin Island they attempted to work northward to the whaling grounds, but on this side as well the ice threw up a solid barrier. It was all the men could do to keep their ships clear of the shifting floes. Finally, in October, the pack closed in for good, trapping eleven whalers in the vicinity of Home Bay on Baffin Island just north of the Arctic Circle. Six hundred men faced the prospect of enduring a winter in the Arctic with supplies meant only for a summer cruise.

Immediately, the captain of one of the trapped vessels, the *Middleton*, placed his crew on short rations. On board the *Viewforth*, iced in nearby, the mate, William Elder, watched with mounting alarm as the hungry sailors aboard his sister ship loudly plotted mutiny. "If we lie here all winter," mused Elder, "whenever their Provisions gets done they will come and demand Provisions from us and then will the horror of such a Proceeding be severely felt for it will end in nothing but man to man for his dear Life. . . ."[7] Before the grumbling broke out into open rebellion, however, the *Middleton* was crushed in the ice and its fifty crewmen dispersed to find berths in one or another of the neighbouring ships.

Not long after, a second vessel, the *Dordon*, began to fill with water and slowly sink. Crew members abandoned ship, taking with them the entire liquor supply, which they quickly consumed on the ice. That evening some of the drunken men lit a fire in the derelict ship's cabin to warm themselves. The flames leapt out of control to the ceiling, setting the decks on fire. By morning the *Dordon* was a charred hulk, burned down to the water line, its crew huddled shaking and bleary-eyed in tents, their clothes roasted right off their backs.[8]

By the end of November four of the ships managed to break free of the pack, leaving five still afloat to face the dark months of winter. The cold grew intense. No one dared to remain on deck for more than a few minutes at a time for fear of frostbite, and the weak fires burning below gave little warmth. "The ice at the top of my bed is at about $\frac{1}{4}$ of an inch thick," William Elder wrote in his journal, "the decks below are all covered with ice."[9] Frozen rigging rattled in the breeze; boards popped with the frost. Ink froze on the nib of a pen and lamp oil had to be thawed at the stoves before it would burn. When the pack closed like a tightening fist around the ships their hulls cracked, the decks heaved and the masts shuddered. But then, miraculously, the wind and current relented, the pressure slackened, the crisis passed.

In these quiet times, when the ice presented no immediate

danger, time passed slowly. Sailors played cards, read, wrote in their diaries, reminisced about happier voyages and contemplated their uncertain future, summed up by one man in his journal: "If the ship goes apparently nothing but death awaits us."[10]

Rations of salt meat, bread, oatmeal and tea were enlivened from time to time by the meat of foxes and polar bears shot on the ice. British seamen knew that the viscera of the bear could be poisonous, but driven by hunger they often ate them anyway, with painful results. "The most of us ate of the liver and heart in the night, which proved unwholesome," reported an officer on one of the ships, "for we were ill with it and had pains in our heads, our faces were red with scarlet, our eyes ready to start from the sockets, and the skin peeled off our bodies in large pieces."[11]

By the beginning of the new year a worse illness was making its appearance. The blotched skin, bleeding gums and loosening teeth were recognized with horror, and on January 14 the first crewman died of scurvy aboard the *Viewforth*. Caused by a lack of vitamin C, this terrible disease inevitably accompanied a diet of salt provisions. The only cure was fresh meat or produce, and merchant-owners did not supply their captains with anti-scorbutics for what were expected to be brief summer cruises.

Ironically, as whaling historian W. Gillies Ross has pointed out, whalers did have the means to aid their stricken comrades.[12] Whale blubber, and even more the skin attached to it, contains vitamin C and could easily have been distributed to the sailors. But whalemen thought it was beneath their dignity to eat these parts of their prey, even when they were starving, and so the disease worked its way inexorably through the crews. Some of the afflicted lay in their frozen berths listening in silent terror to the boards beside them crack with the pressure of the ice. Others lay helpless in the passageways, rolling like logs from side to side as the ships knocked about in the floes.

One vessel, the *William Torr*, drifted out of sight of the others before Christmas. Four years passed before the fate of the vessel and its crew was revealed by a party of Baffin Island Inuit. According to the Inuit, the whalers abandoned the *Torr* when it foundered in the ice and came ashore near Cape Fry. Twenty-two of the crew were so reduced by frostbite and exposure that they could not continue. Remaining at the Inuit camp, they eventually all died. The others set off across the ice in search of another ship they believed to be nearby. They were never seen again.

Meanwhile, at the end of January, two of the four remaining ships worked free of the ice and made their way back across the Atlantic. On one of them, William Elder's ship, the *Viewforth*, only seven men were healthy enough to stand their watch. When the missing ships arrived at Stromness they discovered that a rescue vessel, the *Cove*, was preparing to set off in search of them. Two whaleships remained imprisoned in the ice so the *Cove* went on its way. While it carried out a methodical search along the edge of the ice pack, the wayward vessels, released at last, arrived home. The final death toll was appalling: the entire crew of the *William Torr* missing along the frozen coast of Baffin Island; fourteen men dead on William Elder's ship; twenty-two fatalities on another ship. A total of six ships lost in the ice.

The whaling community hardly had time to absorb the tragic results of the 1835 season when another season of death was upon them. For the second straight year fields of impenetrable ice clogged Davis Strait, and October found six members of the fleet trapped well up in Baffin Bay. During the severe winter that followed, only one of the vessels went down. The rest were held for six months while cold and scurvy turned them into floating coffins. Drifting slowly southward with the pack, the ships were finally released and late in April they began arriving at Stromness, their decks littered with the dead and the dying, their holds stinking with disease. The *Norfolk*–sixteen members of its crew dead. The *Advice*–seven out of forty-nine left alive. The *Dee*–fifteen survivors out of sixty. The *Grenville Bay*–twenty deaths. And the *Swan*, not home until July, whose final death toll no one even bothered to count.

The whalers knew whom to blame for these terrible misfortunes. Experience showed that there would be seasons when the whaleships would not be able to escape from the ice. But merchant shipowners, instead of providing for this eventuality, pretended that their vessels needed only enough supplies to complete a summer voyage. Wishing to rein in their expenses, the penny-pinching owners sent no arctic clothing, no fresh provisions, no anti-scorbutics. Crews caught in the ice were expected to muddle through by hunting game on the floes.

But the decline of Davis Strait whaling that set in after the mid-1830s was due to more than tight-fisted owners. The destruction of so many ships forced several merchants out of the business. It was obvious that the farther the fleet penetrated into the Arctic, the greater the risk that a vessel would not return. In the quarter century between 1819 and 1843, at least eighty-two ships were lost. What is more, the number of whales was declining rapidly. Captains found it harder and harder to fill their holds. Vessels withdrew from the business until by 1841 only one-fifth as many whalers visited Davis Strait as had done so a decade earlier. The British fleet never again numbered more than thirty vessels in a season.[13]

III

As they contemplated the critical situation of their industry, British whalers agreed that the establishment of a permanent station on the shores of Baffin Island would be a step in the right direction. The crew at a shore station would hunt whales on its own, as well as search out new whaling grounds and offer assistance to ships in distress. The scheme received the support of James Clark Ross, probably the most famous British naval explorer of the period, who warned that without some kind of shore station "this best nursery for our seamen, and this important source of national wealth will inevitably be lost to the country."[14]

However, the plan had one important drawback. The coast of Baffin Island was almost entirely unknown. What is now known as Cumberland Sound, for example, was bandied about as an ideal location for the proposed station. But no European had visited the spot in more than three hundred years and, frankly, no whaler was quite sure where it was. Before a station was feasible, the coast had to be explored.

The person who did more than any other to promote the whaling station project was a young captain from the Scottish port of Aberdeen, William Penny. Captain Penny was not a conventional whaling-master. Instead of the grim countenance of a weathered old seadog, his portrait reveals a set of soft, finely chiselled features, framed by a neatly trimmed beard. Decked out in native sealskins he looks more like a matinee idol than the captain of an Arctic sailing ship. While most whalers were content to return to the same hunting grounds each season, Penny was an innovator, always seeking new grounds and the backing to exploit them. He was a plain-spoken man, uncomfortable

Packing whalebone. Once the pieces of baleen were removed from the whale's mouth, they were scrubbed clean and bundled for storage.

in the drawing rooms of polite society and the corridors of power. More than once he saw his plans for improving the fishery frustrated by more cautious, less visionary associates, and he blamed the failure of his projects on his own lack of influence in official quarters. But in the Arctic, where what he knew was more important than whom he knew, his fellow whalers admired him for his zeal, his knowledge and his courage.[15]

The son of a whaling man, Penny shipped on his first voyage when he was just twelve years old. By the time he was master of his own vessel he had made several visits to the Davis Strait grounds and knew from Inuit stories about a deep bay along the southeast coast of Baffin Island where the bowhead still gathered in large numbers. In 1839, on Durban Island, Penny met a young Inuk named Eenoolooapik who had been born on the shores of the bay and was eager to lead the British whalers to it. Eenoolooapik returned with Penny to Aberdeen for the winter where he was dressed up in waistcoat and necktie and shown off at fashionable dinners and theatre parties. Like so many of the northern natives exposed to the moist European climate, Eenoolooapik fell seriously ill with a lung infection and came very close to death. But he recovered, and the next summer was back in Davis Strait with William Penny aboard the whaler *Bon Accord*.

Heavy ice blocked the Melville Bay passage that season, so Penny crossed to the Baffin shore at about latitude 65 degrees. As he came within sight of the high coastline, the mouth of what appeared to be a broad inlet opened in front of him. This, confirmed Eenoolooapik, was his native territory. A triumphant Penny led three other whaleships into the inlet. The explorer John Davis had been the last European to cruise these waters, in 1587, and he was convinced that it was an open passage leading away to the west. The whaleships probed the narrow fjords that indent the shoreline and concluded that Davis was wrong; they were in a large inlet, not a strait. Penny christened it Hogarth's Sound. One of the other whaling captains later produced his own map of the area and called it Northumberland Inlet. Today it is known as Cumberland Sound.

In September, just as the Inuit promised, the whales appeared in large numbers. Though the *Bon Accord* failed to make a kill, the other ships did and Penny sailed home confident that he had discovered an important new whaling ground. The owner of the *Bon Accord* could not have cared less. With nothing to show for the voyage, he was forced to sell his ship, the last of

Boat crew. The crew of a boat aboard a British whaling ship in Davis Strait in 1889.

the once-mighty Aberdeen fleet. Captain Penny was out of a job and, while whalers began to visit Cumberland Sound regularly, the idea of building a whaling station there languished.

While British whalers stalled, their American counterparts once again proved more energetic. Preoccupied by sperm whaling to the south, Americans had abandoned the Davis Strait region following the Revolution, but as sperm whales became scarcer, and the value of baleen whales increased, Yankee vessels returned to the Arctic. It did not take them long to appreciate the value of a wintering establishment on Baffin Island. The bowhead entered Cumberland Sound in the spring long before ships were able to break through the ice. Shore parties that overwintered on the coast could hunt these early arrivals with Inuit help in boats launched from the floe edge. In 1851 Captain William Quayle, of the Connecticut whaler *McLellan*, decided to try an experiment. He left his mate, Sydney Buddington, and eleven sailors to sit out the winter in Cumberland Sound, living in a stone hut covered with sealskins, eating raw seal and walrus meat. Thanks to the local Inuit, the men survived comfortably and in the spring justified the experiment by killing seventeen whales before they were picked up.

When he heard about this incident, William Penny was horrified. Not only had others hijacked his overwintering project, but worse, they were Americans, who had no right to challenge British hegemony in the Arctic. Fired by a potent mixture of patriotism and commercial ambition, Penny formed the Aberdeen Arctic Company and sent out two ships to Cumberland Sound, the first vessels to voluntarily spend a winter in the ice. In the summer of 1854 they emerged with a catch of thirty-nine whales, again thanks largely to Inuit help, and Penny was giddy with grand schemes. He foresaw a permanent colony in the sound with steam-powered ships doing the whaling, miners digging up the island's mineral wealth and missionaries instructing the Inuit. But his partners in the Aberdeen Company were more cautious and Penny fell out with them. Once again he was on his own. Never again did he manage to put together the capital to realize his dream of a Baffin colony. In 1864 he retired from the sea, at about the same time the Aberdeen Arctic Company went out of business.

Penny's personal ambition was frustrated, but his confidence that whalers could spend the winter in the Arctic, either on ships or at shore stations, proved fully justified. From the middle of the nineteenth century, both Americans and British left wintering vessels and established stations at various points along the coast of Baffin Island and also in the northwest corner of Hudson Bay where bowhead were plentiful. Land-based operations had advantages over whaling from ships. Stations did not go down in the ice. They required a smaller complement of men; indeed, most relied almost completely on Inuit labour. And they brought in a variety of trade goods—furs, walrus hides, ivory—to supplement the whale catch.

Overwintering marked the beginning of a new relationship between whalers and Inuit. The Inuit supplied the newcomers

Arctic whalers. The crew of an American Arctic whaleship poses in 1904, near the end of the whaling era in the eastern Arctic.

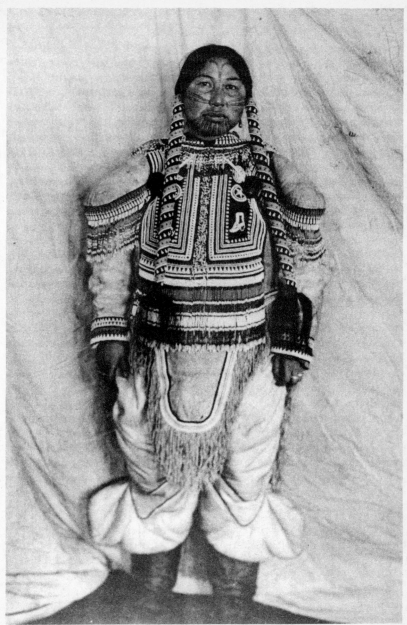

Inuit woman. One of the Canadian Inuit who supplied the Arctic whalers with food and clothing during the long winter months.

with food, trade goods and labour. They were employees now as well as trading partners, interrupting their own annual routines to help the whalers survive and hunt. This new arrangement brought many benefits, chief among them the guns and metal goods that previously were unknown to the Inuit. But there was a price to pay. The white man's diseases were fatal to a people who had built up no immunities to them. During Captain Penny's first winter in Cumberland Sound, cholera killed off an estimated one-third of the local Inuit population. Fifty years later, on Southampton Island in northern Hudson Bay, typhus originating with a whaling station supply vessel wiped out virtually the entire population. It was the same story wherever native and newcomer mingled.

Other effects of contact were less obvious but equally profound. The Inuit supplied the wintering whalemen with stocks of fresh meat, and local animal resources were devastated by overhunting. In return the Inuit took to using manufactured goods such as canvas tents, woollen clothing and canned food. Increasingly, the economic life of many of the northern people became centred on the activities of foreign whalers. Their traditional seasonal cycle of subsistence hunting and fishing was replaced by a new pattern that included periods of wage employment.

Still, the impact of the whaling era on native northerners should not be overstated. Whalers were only seasonal visitors. Many Inuit had nothing to do with them and remained on the periphery of the white man's economic orbit for decades to come. It would take the arrival of missionaries, fur traders and government agents before the Inuit felt the full impact of southern civilization.

IV

By the middle of the nineteenth century declining catches were sending the British whaling industry into depression. Shore whaling was one of the innovations that helped to retard this decline. Just as important were advances in technology that allowed the whalers to chase and kill their prey with new efficiency. Sir John Ross inaugurated the steam age in the Arctic when he outfitted his ship *Victory* with an auxiliary steam engine before setting off in search of the elusive Northwest Passage in 1829. The experiment was a failure. The screw propeller had not yet been invented and the *Victory* was encumbered with paddle wheels. The engine kept breaking down and the boiler leaked badly, even when caulked with a frightful mixture of potatoes and dung. Eventually Ross discarded the machinery on a frozen Arctic beach.

But the *Victory* episode represented a setback, not a defeat, for steam technology. During the 1840s the propeller replaced the awkward paddle wheels and British naval vessels began to convert to steam. Whalers remained dubious, but their doubts were dispelled in 1850 when two screw steamers participating in the search for John Franklin forced their way through Melville Bay ice that stopped sailing vessels in their tracks.

In 1857 whalers installed the first auxiliary engines in their vessels; two years later the first steam whaleships were built from scratch. Not that conversion was ever complete. The engines were small and meant to complement wind power. Ships still carried sails, which were unfurled on voyages to and from the whaling grounds to conserve coal. Nevertheless, in rough seas and heavy ice, steam made a crucial difference. Additional speed allowed whaleships to probe deeper into the Arctic in search of new hunting grounds. Additional power allowed the vessels to endure the buffeting of wind and storm. Increased mobility allowed them to thread their way through the narrowest of open leads or smash through all but the thickest ice barriers. Steam did not tame the Arctic; the Breaking-Up Yard retained its horrors. But the increased safety and mobility provided by steam power were a great help to whaling captains. By the late 1860s almost the entire British fleet had converted.

At about the same time, modern weaponry was transforming the hunt for whales. The technique of chasing the animals in open boats with hand-held harpoons had not really changed since the days of the Basques. Whalers had long hoped for a weapon that would kill their prey instantly, at a safe distance. This need was especially acute in the Arctic where a harpooned whale could run under the ice and escape. Early in the 1850s, C. C. Brand, an armourer from Norwich, Connecticut, invented the first successful bomb-lance. This was the weapon used so effectively by Captain Charles Scammon to slaughter gray whales in the lagoons of Baja, California. Once a harpooned whale was run to exhaustion, the boatheader fired an explosive missile into the animal from a distance. Armed with a time-delay fuse, the missile detonated, killing or mortally wounding the whale. By the 1860s the bomb-lance had pretty well replaced the hand-held lance in the whaler's arsenal.

Taking its place next to the bomb-lance was the darting gun, invented by New Bedford whaling captain Ebenezer Pierce in 1865. This device married the accuracy of the conventional harpoon with the deadliness of the explosive bomb. Ideally, it fastened to the whale and killed it at almost the same instant. As the harpoon plunged into the animal, a slender rod was pushed back and activated the trigger of a gun attached to the shank of the harpoon. The gun fired an explosive into the whale. The force of the explosion knocked the gun into the water, where it was retrieved by the harpooner, while the metal harpoon remained fixed to the whale.

The darting gun was particularly popular with American Arctic whalers. Their British counterparts preferred the harpoon gun, either held at the shoulder or mounted at the bow of the boat. This cannon-like weapon fired the barbed iron with greater velocity and over longer distances than even the strongest harpooner could achieve. Later the harpoon gun would be armed with explosives and installed on speedy, steam-powered catcher boats, ushering in a new era in whaling history. For the time being, however, British bowhead whalers in the eastern Arctic used the harpoon gun with good effect in their traditional wooden boats.

V

The season of 1866 was remarkable for its frustrations. The Arctic fleet arrived off the north end of Baffin Island early in July to find the ice too plentiful for whaling. Thinking only of their own safety, the ships began to pick their way southward through the masses of ice. One by one they made good their escape, until only the *Diana* remained. This 355-ton whaleship from Hull was fitted with an auxiliary steam engine, but it hadn't the power to force a passage through the drifting floes, and by the middle of September Captain Gravill had to make a decision. He could take his ship back to Pond Inlet, secure it for the winter and hope to last out the dark months on whatever food and fuel the frozen country might provide; or he could lodge the *Diana* right in the middle of the pack ice, counting on its southward drift to carry him and his forty-eight men down Davis Strait and out into the Atlantic before they froze or starved to death. Gravill made his decision; on September 22 he committed ship and men to a long winter's drift in the ice. "We are playing our last card for dear life," the *Diana*'s surgeon, Charles Edward Smith, confided to his journal.[16]

Almost immediately the captain placed his crew on short rations. Meals consisted of meagre plates of hard bread, salt meat, oatmeal, potatoes and peas. The stove, fuelled now with barrel staves, spare oars and even the upper masts and yardarms, was lit each morning just long enough to cook the day's hot meal. After that the men made their own warmth as the temperature dropped to minus thirty degrees Celsius in the crew's quarters. Wine froze in the bottle; ice formed on the walls of the cabins, glistening in the dim lamplight. At night the sailors huddled in their berths—"your feet frozen as dead and cold as masses of ice," wrote Smith, "your backbone feeling like one long icicle, your hands tingling with cold, your nose pinched blue, your breath congealing upon the blankets."

Meanwhile, the ice pack carried the *Diana* slowly down the coast of Baffin Island and, at the end of November, into Frobisher Bay where for the next three months it was tossed back and forth at the whim of the current. All this time the crew lived in mortal dread, expecting to be overwhelmed by the ice at any time. The pack was a vise, tightening in wind and storm until, wrote Smith, "the ship [was] bending, writhing, and straining as though she were some living creature struggling in the agonies of dissolution." Then, just as the vessel threatened to burst apart, the pressure slackened, the weather would moderate, and once again the men were preserved. "This is a dreadful life," moaned surgeon Smith. " 'Tis wearing us out far more rapidly than cold or want of food, or any other privations which we are enduring now. The slightest movement of the ice, any unusual noise or bustle upon deck, rouses us up in a moment. We cannot sleep, we cannot even eat what little food we have, we cannot rest below deck, in the cabin, upon deck, anywhere. Restless, uneasy, anxious, we will not have a moment's peace of mind or body so long as we are in this awful ice."

The hull of the *Diana* opened under the strain and water began to fill the hold. Men constantly worked the pumps to keep the ship afloat. The day after Christmas, Captain Gravill died, as much from cold and nervous exhaustion as from disease. Whale flesh now fed the fire, filling the decks with a loathsome stench. The new year brought with it a cold even more intense. "A raven flew over the ship this morning," noted Smith in his journal. "The bird's neck was encircled by a glittering ring of ice formed by the freezing of the moisture in its breath upon the feathers. As the bird approached us, this ring of ice sparkled in the sunlight like a diamond bracelet."

Scurvy appeared early in January. Smith described the symptoms. "They complain of pain and tenderness of the gums, which are swollen, livid and spongy, and bleed to the touch, whilst the hard and soft palates are inflamed and the teeth loosening." Red spots appeared on the legs, then blossomed into patches of deep purple, like bruises. The stricken lay apathetically in their bunks, waiting to die, refusing to respond to the surgeon's urgent appeal that they take some exercise on deck. The small stock of lime juice was quickly exhausted. Smith could feel his own teeth loosening in their sockets as he mixed placebos of harmless chemicals in an attempt to keep up the spirits of the sick and dying. On February 13 the first sailor died. By the end of the month the majority of the crew showed symptoms of disease.

At the beginning of March the current carried the *Diana* out of Frobisher Bay and the ship began moving south again. Soon the lookout spied the coast of Labrador off to the west. Leads began to open in the ice as the pack broke apart. Finally, on March 17, after 175 days trapped in the ice, the vessel steamed clear into the North Atlantic.

But the ordeal was not over. The voyage back to Britain was a race against death. The scurvy did not relax its grip and the men continued to die. Only a handful had the strength to carry out their duties. Milder temperatures melted the ice that had formed on the inside walls of the hull, and berths containing the sick were awash in filthy water. No fuel remained to build a drying fire. One sailor died as he attempted to get up in the morning; his body slumped over two other victims of the disease who were too weak to shift the corpse and lay under it for several hours. By the time the *Diana* reached the Shetland Islands, eight corpses lay wrapped in canvas on the deck. Scurvy had claimed two more victims just hours before, and over the next few days three more crewmen died. In all, more than a quarter of the crew did not survive the terrible whaling voyage of 1866.

VI

Like the unfortunate *Diana*, the British Arctic whaling industry in the latter half of the nineteenth century was gripped by forces that it could not control, carrying it in directions it did not wish to go. Unlike the *Diana*, the bowhead fishery did not survive. It was doomed by the increasing cost of carrying on a business fraught with high risks and the near-total disappearance of the animals upon which it depended.

From the 1830s, whalers were fighting a losing battle against the inexorable decline in the number of bowheads. By utilizing shore stations and adopting steam power and new weapons, the British fleet squeezed sufficient returns from the Davis Strait fishery to keep it profitable. As well, whalers began combining their usual operations with the spring seal hunt on the ice off Newfoundland and Labrador. Without the profits from sealskins and seal oil, many whalers would have abandoned the hunt long before they eventually did.

The demand for whale products remained strong. With the discovery of petroleum, the future of whale oil as a lighting fuel looked bleak, but new markets emerged. In the Scottish whaling port of Dundee, for example, oil was used to treat imported Indian jute before it was spun into sacking and other products. Increasingly, the ups and downs of the oil market were far less important than the spectacular increase in the value of baleen. Useful as it was in the manufacture of a variety of products, baleen found a huge demand in the women's garment industry as a passion for wasp waists and full skirts swept the salons of Europe and North America. Strong and flexible, whalebone was the perfect material for the corsets that bound women's figures and the hoops that flared their gowns. In the latter half of the nineteenth century the value of baleen skyrocketed in Britain from around £500 per ton in the 1870s to £3,000 per ton in 1902.[17] Similarly, in the United States the price climbed from thirty-two cents a pound to more than five dollars.[18] It is estimated that by 1890 whalebone constituted 83 percent of the value of a bowhead whale.[19] A single, good-sized whale was worth between $10,000 and $15,000, enough in itself to underwrite the expenses of a whaling voyage.

The price of bone fell off after 1907 as fashions once again changed, but by that time Arctic whaling was in its death throes anyway. The industry collapsed not because of a fall in demand, but because of a failure of supply. By the end of the century there simply were not enough bowhead left to sustain the hunt. In the summers of 1907 and 1908, Captain Joseph-Elzéar Bernier toured the Davis Strait area on a patrol for the Canadian

Whaling station. This American station just north of Frobisher Bay was typical of several established along the east coast of Baffin Island at the tail-end of the Arctic whaling era.

government. Where once a fleet of 150 sailing ships had pursued the bowhead, Bernier counted just 8 vessels. Where once a single whaler had captured 34 animals in a few weeks, Bernier sighted a lone whale in two summers of cruising. "It must, therefore, be admitted," he reported, "that, at least for the present, the whaling fishery is exhausted."[20] Before the outbreak of the First World War the truth of Bernier's observation was admitted, and the last British whaleships sailed home from Davis Strait, never to return.

For the bowhead, it may have been too late. The international community banned the killing of bowhead whales by anyone other than native northerners in 1931, but recent assessments of the population size in the eastern Arctic are not encouraging. In the entire region—that is, northern Hudson Bay, Davis Strait and Baffin Bay—there are thought to be no more than a few hundred bowhead remaining.[21] In other words, fewer of these whales may be alive today than were slaughtered in one summer by whalers early in the nineteenth century. Still not known is whether these animals represent a population on its way back from near extinction, or the dwindling remnant of a doomed species too depleted to recover.

THE LAST FRONTIER

*There is heavy responsibility resting upon the master
who shall dare cruise different from the known grounds,
as it will not only be his death stroke if he does not
succeed, but the whole of his officers and crew will unite
to put him down.*[1]

–Captain Thomas Roys

Whaling resembled the gold rushes that swept across western
North America in the nineteenth century. Like prospectors
restlessly searching for the big strike, whaling captains moved
from one hunting ground to another, swiftly killing off the
animals in one place, then passing on to the next. Driven with
the same sense of urgency as the gold-seeker, whalers repeatedly
sailed off the maps of the known world, seeking new sources of
wealth before the old grounds played out.

Thomas Roys was one of the sea-going "prospectors" who
ventured out ahead of the pack. In 1848 he decided to follow up
rumours that the Bering Strait, the narrow passage that separates Siberia and Alaska, was the gateway to a rich hunting
ground teeming with a new species of "polar whale." The
occasional explorer had sailed through the strait, but no whaler
had ventured into the icy seas on the other side. So dangerous
were these waters considered that Roys had to trick a crew into
coming with him. But his path-breaking voyage sparked a
whaling "rush" into the western Arctic that lasted fifty years and
provided a moribund American whaling industry with its final
flurry.

I

The men aboard the *Superior* were terrified. They had signed on the 265-ton, three-masted bark at Long Island's Sag Harbor for a ten-month whaling voyage into the South Atlantic. Yet here it was a year later, the end of July 1848, and they found themselves at the opposite end of the world, drifting in a thick fog through Bering Strait toward the unknown horrors of the Arctic Ocean. When told where they were going, the first mate began crying for the home he expected never to see again and the rest of the crew came close to mutiny trying to convince their captain to turn back.

But Thomas Roys was no ordinary whaling-master.[2] For fifteen years he had sailed the world's whaling grounds, always returning with a full hold. His success was taking on the character of legend and he was only thirty-two years old. It was a green sailor who had not heard how Roys had actually ridden on the back of a whale. The captain was chasing a pod of right whales in the North Pacific, so the story went, when he became entangled in a line and was pulled overboard. While he struggled in the water, one of the animals rose directly underneath him and Roys found himself sitting on its back, much like a bull rider at a rodeo. Trying to jump from the whale, Roys landed on the back of another, then fell into the water where the flukes of a third slapped him back into his boat. Unruffled by this close call, he picked up a lance and smoothly killed three of the animals.

Roys had made a study of the "polar whale." Listening to the accounts of seamen who had sailed through Bering Strait and reading the journals of explorers, he became convinced that a large population of black whales lived unmolested in the Arctic, waiting for the first whaleship bold enough to come looking for them. Realizing that no shipowner would send him on such a risky mission, Roys pretended to be taking the *Superior* on a brief Atlantic cruise. But once at sea he sailed his vessel all the way to Tasmania before writing to his employers to admit that he was heading for the Arctic. By the time they received the letter it was, of course, too late to stop him.

When the fog at last lifted, the crewmen of the *Superior* found themselves surrounded by spouting whales. Were they a herd of familiar humpbacks, as the mate alleged, or were they "the new-fangled monsters" their captain had told them about? Roys ordered the boats away to find out. Still unsure what to expect, the sailors approached the animals warily, but Roys was delighted to find them slow-swimming and placid. He was first to the kill, planting two harpoons in the back of a large specimen. Immediately it dived and remained underwater for fifty minutes. "I began to think that I was fast to something that breathed water instead of air," wrote Roys later, "and might remain down a week if he liked." But the whale came to the surface at last and Roys lanced it without incident, making the first commercial whale kill north of the Aleutian Islands.

Once they began cutting in, the officers admitted that this was a whale unlike any they had seen before. Its baleen measured an incredible twelve feet in length and the blubber yielded 120 barrels of oil, a huge amount for the size of the animal. In the Davis Strait region, European whalers had been hunting these whales for more than a century. But Roys and his men were veterans of the southern fishery and had never seen a bowhead. Their voyage confirmed that this species of whale inhabited the waters of the western Arctic as well as the straits and inlets at the eastern end of the frozen archipelago.

The *Superior* remained north of Bering Strait for a month. Much to the relief of the crew, the weather turned out to be quite mild and the ice easily navigable. The ship took eleven whales, producing enough oil to fill the hold, then sailed back through Bering Strait. Early in October Roys reached Honolulu where he began spreading word of his discovery among Yankee whalers growing discouraged at the decline of the sperm whale hunt.

The rush was on. During the summer of 1849, fifty vessels followed the *Superior*'s lead into the Arctic and almost three times as many arrived the next season. These newcomers found herds of fat whales that exceeded even their fondest expectations. Animals were killed that yielded an incredible three hundred barrels of oil, worth more than $4,500 at 1850 prices, and that was before counting the value of the baleen. One ship tryed out 4,200 barrels of oil while cruising the northern waters.[3] The western Arctic appeared to be an El Dorado for whalers.

As for Thomas Roys, he made one more voyage to the Bering Strait area, spending three summers in the north and returning at last to Long Island with a cargo worth $100,000. Then, true to his nature as a restless innovator, Roys abandoned

151

the lucrative bowhead hunt to begin experimenting with new weapons for killing new species of whales.

II

The bowhead of the western Arctic pass the winter in the Bering Sea along the southern edge of the solid ice pack that extends down through Bering Strait from the north. In April the ice begins to break up, opening lanes of clear water through which the whales migrate northward. Feeding on the swarms of tiny organisms living near the surface of the water at the ice edge, bowhead cruise slowly through Bering Strait into the Chukchi Sea where some swing to the west along the coast of Siberia while others continue around the top of Alaska into the Beaufort Sea. In the fall, when the ice once again moves down from the north, the whales beat a retreat back the way they have come, departing the Arctic through Bering Strait in October and November.

After outfitting at one of the Hawaiian ports in March, an Arctic whaler tried to be skirting the edge of the Bering Sea ice pack by mid-April. The ice presented fewer perils here than in the eastern Arctic, but the pack frequently closed around the fleet, holding it immobile for days on end while snowy gales lashed out of the north and thick fogs made it impossible to see one end of a ship from the other. By the end of June the whalers penetrated Bering Strait and began working their way north along the coast of Alaska toward Point Barrow. This part of the voyage resembled the creep across the Breaking-Up Yard in the eastern Arctic. A shift in the wind might drive the ice down upon the ships, crushing them helplessly against the shore. Most of the whales were ahead of the fleet at this point. The hunters fell in with their prey again in September and October as the animals returned south. As the sea ice re-formed and autumn gales battered the ships, the fleet took whales as long as it safely could, then hurried through the strait and on to Hawaii to regroup for winter whaling in warmer climes.

The easy hunting that greeted the first whalers in the western Arctic did not last long. Four seasons after Thomas Roys announced to the world the existence of the bowhead herd, 220 Yankee whaling ships crowded onto the new grounds. Almost immediately the catch declined. As the animals became more scarce, so did the whalers. Then, bad as the situation was, it got worse.

Just before noon on June 22, 1865, the men of the New Bedford vessel *William Thompson* were busily boiling oil in the shadow of Cape Navarin off the coast of Siberia near the top of the Bering Sea. When they saw the sails of a ship emerge out of the fog from the south, the sailors probably thought it was just another whaler come to join the several dozen present in the North that summer. But the newcomer was no whaler. It was the Confederate raider *Shenandoah*, bringing the American Civil War to the Arctic's doorstep.[4]

As the *Shenandoah*, bristling cannon from every porthole, descended on the whaleship and its stunned crew, the Civil War was in fact already over. But Captain James Waddell did not know it, and if he suspected, well, he kept his suspicions to himself. Waddell, an officer in the Confederate navy, had set out from England eight months before, determined to strike a blow at the economy of the Northern states by hunting down their whaling fleet. A converted merchant vessel, the *Shenandoah* was a huge ship, at 1,100 tons at least twice as large as the biggest whaler, measuring 222 feet in length and powered by an 850-horsepower steam engine as well as sails. Waddell took this speedy titan around the Cape of Good Hope to Australia, capturing whatever Union vessels he could along the way. At Melbourne he took on the rest of his crew, in violation of British neutrality, then continued on his way through the Pacific towards a rendezvous with the Arctic whaling fleet.

The crew of the *William Thompson* was powerless to stop the Confederate boarding party as it came streaming over the sides to take control of the whaleship. They passively accepted imprisonment aboard the *Shenandoah*, and watched as the raiders set fire to their vessel and another captured nearby. The next day Waddell ran down four more ships, three whalers and a trading brig. Designating one a prison ship to carry the captured crewmen south, he set fire to the other three, then disappeared into the fog, heading north. In light winds, the whalers were unable to outrun the steam-driven raider and on the night of June 25–26 Waddell captured and torched another four vessels. His decks were so overcrowded with prisoners that he put them in whaleboats and towed them along in his wake. "It was a singular scene upon which we now looked out," remarked Cornelius Hunt, mate aboard the *Shenandoah*. "Behind us were three blazing ships, wildly drifting amid gigantic fragments of ice; close astern were the twelve whale-boats with their living freight; and ahead of us the five other vessels, now evidently aware of their danger, but seeing no avenue of escape." Three of these five fell prey to the marauding raider.

The whaleships were helpless to resist the well-armed Confederates. A few hand-weapons and their whaling tools were all the whalers had with which to arm themselves. One captain remembered that he had an old cannon below decks. Breaking it out, he gave it a test firing. "When the gun went off it rose right up in the air, coming down on the deck, made a great hole on the planking. Everybody was scared half out of their wits, while the concussion was so great that it broke all the glass in the skylight to the cabin." Firepower of this calibre was not going to stop the *Shenandoah*.

Waddell next fell in with a squadron of ten whalers in Bering Strait conveniently gathered around one that had struck a piece of ice the day before and was sinking. Realizing that they could not escape in a flat calm, the whalers surrendered. The one exception was seventy-year-old Captain Thomas Young who refused to give up without a fight. Grabbing a loaded harpoon gun, he prepared to fire on the *Shenandoah*'s boarding party, but the weapon had been disarmed by his crew who didn't want to see the old man hurt and he gave up without further resistance. Most of the ships were burned to the waterline; two were sent south loaded with prisoners.

At this point, as the rest of the whaling fleet dispersed in a panic, Waddell abruptly turned around and left the Arctic. Apparently the Confederate captain had learned from newspapers on the captured whalers that the war was moving rapidly to a conclusion, that the South was disintegrating, and that, without the sanction of war, he was nothing more than a pirate. Knowing that the United States navy would be looking for him, Waddell slipped furtively back to England without putting in at any port. Had it still been wartime the voyage of the *Shenandoah* would have been hailed as a great triumph for the Confederacy. Waddell had sailed completely around the world, captured over a thousand prisoners and destroyed thirty-four ships worth over one million dollars, all without taking so much as a single shot from the enemy. But it was not wartime. The *Shenandoah* was an outlaw and an embarrassment. Americans demanded reparations for the ships it sank. The issue eventually was settled by an international tribunal that required Britain to pay the United States $15.5 million in gold for damages inflicted by the *Shenandoah* and two other British-built Confederate raiders. Claims were still being paid out to the owners, insurers and crewmen of the sunken whaleships well into the 1880s.

The cruise of the *Shenandoah* was not the only blow sustained by the whaling industry during the Civil War. In the Atlantic the *Alabama*, a second armed Confederate cruiser, destroyed a dozen Yankee whalers. The Union government conscripted another forty of the older hulls into the "Great Stone Fleet," vessels it purchased, filled with stones and sank in Charleston harbour to impede Confederate shipping. Whale merchants did not replace any of these lost ships. Quite the reverse, they withdrew even more of their vessels from a business that was proving increasingly risky and profitless. By the end of the war the American whaling fleet was half the size it had been at the beginning.

Given the decline in whale stocks and in the demand for animal oil, old-style whaling was probably doomed anyway. But any chance it had for recovery following the setbacks of the war was frustrated during the 1870s by two of the worst shipping disasters in the history of the Arctic.

III

During the summer of 1871 the ships of the Arctic fleet crept steadily northward along the shoal-ridden coast of Alaska.[5] If wind and weather conformed to the usual pattern, the fleet could expect to follow open leads next to the mainland all the way to Point Barrow. But in the Arctic nothing is predictable, and at the beginning of September the ice pack pressed in around the ships, trapping them against the shore.

The first to go down was the *Roman*, anchored in shallow water off Point Franklin. The oncoming ice smashed in the stern of the vessel, then piled up against the hull, driving it over on its side and opening a gaping hole. The crew escaped by hauling three boats across the surging ice to open water, then sailing twenty miles to the nearest ship. Early the next morning the ice overwhelmed the *Comet*; its crew likewise got away in boats to safety.

A week passed. Strung out in a thin line along the coast, the whalers waited for some break in the weather, for a favourable wind to shift the ice and relieve the pressure. But the ice continued to press down upon them. Finally the captains agreed that if they did not wish to be trapped for the winter the time had come to abandon the fleet. Seven ships, they knew, had escaped the ice to the south and were waiting to assist the trapped vessels. "Tell them all," one of these captains sent word, "I will wait for them as long as I have an anchor left or a spar to carry a sail."

To reach their rescuers, the 1,219 men, women and children aboard the thirty-two trapped whaleships had to cross sixty miles of churning ice through worsening weather in overloaded open boats. They prepared for their exodus methodically, raising the gunwales of the boats for heavy weather, sheathing them in copper and putting on false keels to protect their thin cedar planking from the wear and tear of the ice. Once provisions were loaded and crewmen had bid a sad farewell to ships and shipmates, each set of boats slipped away to join the swelling flotilla moving determinedly southwards down the coast.

When possible, sails were set and the boats swept across stretches of open water, trying to avoid the treacherous masses

Reaching safety. The crews that had to abandon their vessels in heavy ice off Alaska in 1871 reach the rescue ships.

of floating ice. When the wind failed, men took up their oars. Night fell and the ghostly shapes of abandoned ships, heaved over on their sides, emerged from the mist as the boats passed, then disappeared back into the inky blackness. A strong wind came up, and with it a cold rain. Fires dotted the shoreline where fugitives had paused briefly to take a meal and warm themselves. The murmur of voices, the splash of oars, came faintly across the water.

For the most part, the flotilla ran through the narrow channels close to shore. When ice blocked the way, passengers disembarked and hauled the heavy boats across to the next channel. A few miles from the rescue ships the refugees broke free of the ice into open water, exchanging the calm of the inside passages for the rough seas of the outer coast. Tall waves swept over the boats, threatening to swamp them just when they seemed so close to safety. The passengers huddled together for warmth and courage, their clothing caked with frozen spray, until, boatload by boatload, they arrived at the rescue vessels and were plucked from the churning sea.

After five days the desperate evacuation was complete, and the rescue squadron sailed away home. Not a single life was lost. For that the whalers were thankful, but for little else. Thirty-three vessels went down in the ice that season. Shipowners despaired, insurance companies inflated their rates, and the industry journal, *Whalemen's Shipping List*, reported that "There are those who think that the Arctic whaling will be given up in a few years because of the peril attendant on whaling there. . . ."[6]

In 1876, the year of the second great disaster, fewer ships were lost, but many lives. The Arctic fleet had dwindled to just twenty vessels, and half of these found themselves at mid-August trapped in the ice near Point Barrow. Instead of forcing the ships up on the shore, the moving pack carried them around the point towards the Beaufort Sea. As the shoreline receded in the distance, the captains decided to launch the whaleboats and escape to the mainland before it was too late. More than fifty crewmen elected to remain on board. The three hundred others took two days to cross the ice to shore, then travelled south to Cape Smythe, about fifteen miles beyond Point Barrow. Without hope of rescue, the whalers were preparing to spend the winter on the frozen coast when unexpectedly the ice moved away, releasing three ships that had been trapped farther along the coast

The American whaling bark Black Eagle *caught in the ice.*

and that carried the whalers to safety. Only one of the ten abandoned vessels ever reappeared. It was found the following summer and towed back to San Francisco. Inuit reported that five of the seamen who stayed behind eventually arrived at Point Barrow; three survived the winter. As for the rest, they were never seen again.

IV

The awful decade of the 1870s virtually destroyed New Bedford as a whaling port. Too many of its ships lay in pieces along the Arctic coast for the town's merchants to continue risking their capital in such an unpredictable enterprise. What was true of New Bedford was equally true of New England generally. Recalling how just a few years earlier a dozen busy whaling ports dotted the East Coast, one industry newspaper lamented: "Now these ports are silent and deserted; their once busy wharves are vacant and fallen into decay; their streets are grass-grown, and most of the inhabitants have long since departed."[7] For New Bedford, prospects were not so dim, since the cotton textile industry began to thrive at about this time. But it was forced to surrender the crown it had worn for over half a century as America's leading whaling port.

The new centre of the industry was San Francisco. Ever since gold was discovered in California in 1848, whaling captains had avoided this west coast port like the plague. Seduced by dreams of instant riches in the gold fields, sailors deserted their berths in droves, and no captain putting into San Francisco expected to come out again with a full crew. But during the 1860s the Hawaiian Islands lost their popularity with whalers. Port charges at the islands escalated, while the Pacific sperm whale fishery declined. At the same time, the transcontinental railway reached San Francisco, allowing whale merchants to ship their products across the country with speed and economy. By the end of the 1870s what was left of the American whale fleet had shifted its business to the California port and whaling's long association with Hawaii ended.

The pre-eminence of San Francisco was assured by the introduction of steam power to the American fleet in the 1880s, several years after its adoption by the Davis Strait fleet. More expensive than sailing vessels to build and operate, steamers used their increased speed and manoeuvrability to track down

The Mary and Helen. *The first steam-powered whaleship built in the United States. The painting is by Charles S. Raleigh.*

Arctic whales where wind-powered ships could not go, thus increasing the catch and offsetting the additional cost. This, at any rate, was the hope, a hope fulfilled by the first steamer in the western Arctic. Built in Maine for a New Bedford merchant, the *Mary and Helen* measured 138 feet in length and carried a 90-horsepower, single-cylinder engine. After its first voyage to the Arctic in 1880, it returned home with an astonishing catch of twenty-seven whales and a cargo worth $100,000, more than the cost of building and outfitting the vessel.[8]

The success of the *Mary and Helen* prompted a group of West Coast entrepreneurs to build their own steamer fleet and, in 1883, to form the Pacific Steam Whaling Company to pursue the Arctic hunt. The industry entered its final decades buoyed by the rising demand for baleen and powered by steam.

V

In the first days of August 1889, a squad of whaleships dropped anchor in Harrison Bay on the north coast of Alaska. This was as far as they dared venture into the uncharted Beaufort Sea. Soon, offshore, bowhead whales would be migrating past and the whalers were waiting for them. They were also waiting for Little Joe Tuckfield. Tuckfield, a boat steerer at Charlie Brower's Point Barrow shore-whaling station, had taken a whaleboat the previous autumn deep into the Beaufort to the mouth of the Mackenzie River to investigate Inuit claims that the area teemed with whales. Whaling men called the Beaufort "the forbidden sea." It was accessible to them only down a long corridor of water that opened each summer as the ice pack receded from the top of Alaska and the Yukon. The passage was shallow and scattered with dangerous reefs and low gravel islands. The shore offered no safe harbours, and at any moment a shifting wind might drive the ice southward, grinding a ship against the rocky beach. "One of the greatest dangers in Arctic whaling is this going east of Point Barrow," wrote a visitor to the region.[9] Sailing vessels were especially wary, but even in the age of steam the Beaufort remained a closed sea.

When Joe Tuckfield kept his rendezvous with the ships in Harrison Bay he reported that, just as the Inuit had said, the water to the east was "thick as bees" with whales. And when the captains saw his small boat crammed with baleen and heard him describe what sounded like an ideal harbour on the south side of

Herschel Island, they decided to take a look for themselves. Sticking nervously together, seven steamers advanced down the coast into Canadian waters near the Mackenzie Delta where they soon realized that a new hunting ground had been discovered. Whales were numerous and the area was free of ice. At Herschel Island, seventy-five miles west of the Mackenzie, a natural harbour, christened Pauline Cove, provided refuge from ice and storm. Along the coast driftwood piled up in tangled heaps, so there would be no shortage of fuel, and the local Inuit reported an abundance of wild game. After staying just a few days, the whalers hurried back to San Francisco with news of the discovery.

The Beaufort was so distant from San Francisco that whalers wasted most of their season travelling to and from the grounds, leaving little time for hunting. As a result, a new strategy evolved. The next summer, 1890, two ships arrived at Herschel Island prepared to spend the winter so that they would be ready for whaling as soon as the ice dispersed in the spring. The whalers were still nervous about the Beaufort and the pioneer vessels were the smallest available, to better navigate the shallow arctic coastline. The 150-ton *Mary D Hume* was the lightest steamer in the fleet. It was joined by the 250-ton *Grampus* and a small sailing vessel, the *Nicoline*. After an uneventful winter in the ice, this tiny fleet emerged from the harbour in July to begin hunting whales. The *Nicoline* left for home without taking a single animal, but the two steamers met with unprecedented success. Cruising eastward past the mouth of the Mackenzie River as far as Cape Bathurst, the *Mary D Hume* took twenty-seven whales, a full cargo, before it returned to Herschel to rendezvous with its sister vessel. The *Grampus* had taken twenty-one bowhead, and laden with its own bone and most of the *Hume*'s catch it left for the south while the *Hume* stayed behind for another season's whaling. By the time it returned to San Francisco in the fall of 1892 with $400,000 worth of whalebone to its credit, the stampede of whaleships into the Beaufort was under way.[10]

Incoming ships from San Francisco did not arrive at the Beaufort grounds until mid-August, which only left them a month of whaling before young ice began to form on the water overnight, warning the captains that it was time to seek the refuge of their winter harbour and begin preparing for the

Mouthful of baleen. Strips of baleen hang like a curtain from the upper jawbone of a bowhead whale.

frozen months to come. At Pauline Cove, the whalers first collected driftwood in great piles to fuel the shipboard stoves. Once ice had formed around their vessels to a good thickness, they turned to weatherproofing the hulls, covering the decks with board or canvas and packing the sides with a thick wall of snow. Stoves kept the cabins and forecastle at a comfortable temperature, no matter how cold it was outside, and a steady supply of melted ice was available for washing and drinking. In case of fire, a hole was broken in the ice next to the gangway and buckets were piled nearby for use in an emergency.

In the extreme cold of a northern winter the men discarded their southern clothing in favour of Inuit skins. "This consisted of a fawnskin shirt next to the body, with the hair inside," explained Hartson Bodfish, a veteran of several winters at Herschel Island, "knee-length underpants of the same material, and deerskin stockings that came to the knees. When we were aboard the ship this was all the clothing we wore. When outdoors we put on another suit of clothing, also of deerskins, with the hair outside, and boots...."[11] The whalers traded for this clothing with Alaskan natives on the inward voyage, or with the

162

Inuit who settled in snowhouses around the ships at Herschel each winter.

The Inuit, along with their neighbours, the Loucheux Indians, also hunted caribou for the whalers in the foothills and mountains on the mainland. The crew of a single whaleship might devour nine tons of meat in a winter, and anywhere up to fifteen ships occupied the harbour in any one season. Whatever provisions the natives brought in they were paid for at a rate of five or six cents per pound, usually taken in the form of trade goods. The meat was stored in underground icehouses dug in the slope behind the harbour.

The whaling men found a variety of ways to while away the dreary winter days of confinement. Baseball and football games took place daily on the harbour's frozen surface. A nearby hill became a toboggan run and the Inuit loaned their dogs and sleds for excursions to the mainland. The less venturesome preferred card games, or a game of snooker at the billiard table brought from San Francisco and installed in a warehouse on shore. Each year a few captains brought their wives along and the ladies added a touch of gentility to the busy round of dinners and dances.

Whaleships in winter quarters at Herschel Island, 1893-94. With their ships locked in ice at Pauline Cove, whalers relieve their boredom playing soccer and baseball on the frozen harbour. The buildings on the beach at the left are Inuit huts and company warehouses. The painting is by whaler John Bertonccini.

163

But life at Herschel Island was not completely tranquil and law-abiding. For nine months of the year, Pauline Cove was a northern boomtown inhabited by a polyglot mixture of rough seamen. The boredom of enforced idleness, aggravated by resentment at the hard living conditions, fuelled with liquor and provoked by the presence of Inuit women, inevitably erupted into violence. The result prompted one visiting missionary to declare that "the scenes of riotous drunkenness and lust which this island has witnessed have probably rarely been surpassed."

Part of the problem was the poor quality of seaman that American whaling was attracting in its final years. Many members of the crews had never before been to sea. Some were recruited reeling drunk off the streets of San Francisco by boarding-house keepers who received a bounty for each green hand they produced. Others signed on for a free ride to Alaska where they intended to jump ship and strike out across the mountains for the Klondike goldfields. As a result, desertion was a chronic problem. Even though Herschel was several hundred miles from the gold rush, and most sailors knew nothing about surviving a trek across an unknown wilderness in the middle of winter, a few jumped ship each year. They had to be pursued because they often stole valuable food supplies and because their absence left the vessels shorthanded for the coming season of whaling. On rare occasions deserters actually made it across the mountains to Dawson. More often they were brought back to the harbour, frost-bitten, starving, in shackles.

In the early years, liquor flowed freely at Herschel. The Pacific Steam Whaling Company tried to impose prohibition but it could not stop a brisk trade from developing. "Each year a vessel is loaded at and despatched from San Francisco with supplies for this fleet, of which cargo liquor forms a large share," reported a North West Mounted Police officer who visited the island in 1896. "This liquor is sold or traded to the native for furs, walrus ivory, bone, and their young girls who are purchased by the officers of the ships for their own foul purposes."[12] What the whalers did not trade, they drank themselves. And then they fought and even murdered each other. In response, the captains wielded a harsh and unregulated discipline. "It was agreed that officers should be severe when offences and defiance became too serious," wrote one returning crewman, "and keep their mouths shut when they went South. So I am not telling all."[13]

If they did not bring their wives, whaling captains commonly set up housekeeping on the island with local Inuit women. Many of these relationships were disruptive of the native community. Women were purchased with trade goods and liquor, and venereal diseases spread. One notorious captain had such an appetite for young Inuit girls that his ship was called "The Kindergarten." He "rented" as many as five at a time from their parents in Alaska, signing them on as members of the crew. On the other hand, many relationships showed all the stability of recognized marriages, the officers living ashore with the same women winter after winter. Children of these common-law arrangements were sometimes taken south by their fathers to be educated in American schools.

In the end, sex and violence were not the most important legacies of the whalers in the western Arctic. Far more important was the impact of their technology on the Inuit who suddenly had access to guns, boats, manufactured clothing and metalware. The introduction of firearms allowed hunters to slaughter huge numbers of caribou and musk-oxen. These animals were seriously depleted, and as they disappeared the Inuit came to rely more on the whalers for processed food, tents and clothing.

The whalemen also altered traditional Inuit society by introducing elements of the wage economy. Sometimes a departing whaler left a whaleboat and gear with a native hunter who used them to chase whales for pay while the ship was absent. Other Inuit who did not work directly for the whalers at least traded with them. They manufactured leather coats for sale to the southerners and harvested fur-bearing animals as a cash crop. Increasingly the seasonal cycle of subsistence hunting and fishing was interrupted by periods of wage employment. These changes became more pronounced as time wore on, but it was the whalers who initiated the contacts that have had such a profound impact on the Inuit.

Despite the lurid tales of drunkenness and debauchery, life at Herschel was probably not as depraved and violent as missionaries and others claimed. Historian John Bockstoce, the authority on this period, has sensibly pointed out that whaling captains would not frequent a harbour where crewmen were liable to be infected with venereal disease, murdered in a drunken brawl, or seduced into desertion.[14] All these things did occur, but the majority of the seamen passed the winter in sober monotony, as Hartson Bodfish admitted in a letter to his mother describing his daily routine: "Get up in the morning for breakfast, lay around till lunch, lay around till supper, lay around till bedtime, varied by playing cards and setting the men to hauling and sawing wood and once in a while going after a load of ice or meat."[15]

It was with great relief, then, that in May the sailors recognized the first signs of approaching spring. Flocks of geese and ducks heralded the return of warmer weather, and patches of ground emerged from beneath the cover of snow. On the mainland, rivers and streams opened with a rush, while out on the sea the frozen surface turned slushy with meltwater. Using large saws, the whalers cut a passage for their ships out of Pauline Cove, and by the second week of July they were steaming through the open leads towards the eastern hunting grounds.

VI

The presence of so many American whalers in the Beaufort Sea attracted the attention of church and government authorities in southern Canada. In 1893 the Anglican Church Missionary Society asked its missionary at Fort Macpherson to travel across to Herschel Island to investigate claims that the Inuit were being corrupted. Isaac Stringer arrived at the island by dogsled on the first day of May 1893 as the whalers were hurrying to prepare their ships for breakup. Despite the fact that he came to pry into their affairs, the captains welcomed the twenty-seven-year-old Stringer, offered him a hot bath, a berth to sleep in, and the use of the newly-constructed billiard room for his religious services. They were less cordial when the missionary raised the issue of the whisky trade. But Stringer was determined, and at length persuaded the captains to sign an undertaking to stop the sale of liquor to the Inuit. This piece of paper represented an important victory for Stringer, though it was not always enforced.

In the fall of 1897 Stringer moved onto the island permanently, with his new wife, whom he had gone south to marry the year before. The couple lived in a sod house purchased for thirty dollars from the whalers, decorated with window curtains fashioned from Mrs Stringer's wedding gown and heated by a stove made of tin cans. Not surprisingly, the Inuit were indifferent to the Christian message Stringer brought to them. They had their own religious beliefs and their own religious leaders. Stringer persevered, but by his own admission he made little headway and in 1901 he left the island for a new posting. His successor, Charles Whittaker, was either more effective or simply less realistic; he claimed to make great progress converting the Inuit to Christianity. Whether or not this was true, the missionary presence did help to smooth some of the rough edges of life at the harbour and to improve relations between Inuit and whaler.

Politicians in Canada's capital were less interested in Inuit salvation than in the nationality of the whalers and the value of the business they were conducting. Canadians feared to see the American flag flying over territory they considered their own. "If the Americans are permitted to skirt our Western possessions, for Heaven's sake do not allow them to skirt us all around," pleaded Senator W. C. Edwards in a letter to Prime Minister Wilfrid Laurier. "They are south of us for the entire width of our country; they block our natural and best possible outlet to the Atlantic; they skirt us for hundreds of miles on the Pacific and control the entrance to a vast portion of our territory, and the next move if we do not look sharply after our interests, will be to surround us on the North."[16] Laurier shared this suspicion of American expansionism in the Arctic, though he did not want to irritate the United States by plainly saying so. Instead he chose to install police outposts at strategic locations across the North and to "quietly assume jurisdiction in all directions."[17]

In the summer of 1903 two police officers came to Herschel Island, charged with the job of keeping the peace, collecting customs fees on trade goods and controlling the liquor trade. But if Herschel ever had been the Sodom of the North, it certainly was not by the time these Canadian lawmen arrived. Missionaries had already curbed the worst excesses of the liquor traffic, and most of the whaleships were wintering farther to the east anyway. Herschel's halcyon days were over. The police stayed in the North to regulate the trappers and traders doing business there and to show that Canada was serious about its claim to ownership. But whaling had entered on its final days.

VII

Whaling in the Beaufort Sea followed the familiar pattern of boom and bust that characterized the industry around the world. The first ships to arrive found the killing easy and filled their holds to bursting with baleen and oil. As more whalers flocked to the site, the animals became scarcer and scarcer until it was no longer profitable to continue the hunt and the ground was abandoned. In the Beaufort, and the western Arctic generally, whalers pretty well exhausted the resource by the turn of the century. The pre-whaling bowhead population has been estimated at 30,000 animals. American ship whalers killed about 20,000. Several thousand more died at the hands of shore-based hunters. Today there are probably some 7,000 or more bowhead surviving in the western Arctic, many more than inhabit the waters at the eastern end of the archipelago but still just a fraction of the herd that once roamed the northern ice.[18]

The Beaufort Sea hunt was the American whaling fleet's last hurrah. Oil had been declining in value for years. Indeed, many captains simply extracted the whalebone from a dead animal, then cast the carcass adrift without bothering to flense it. Then, in 1908 the bottom fell out of the baleen market as well. The Pacific Steam Whaling Company withdrew its ships and the few vessels still coming north gave up any pretense of whaling, concentrating their efforts instead on trading with the Inuit.

For 150 years American blubber ships had wandered the world's oceans in search of whales. They pioneered the deep-sea hunt for sperm whales; re-invented the on-board tryworks and other novelties that revolutionized the industry; wintered on the frozen Arctic coast for the first time; brought the uncertain benefits of Western civilization to the islands of the Pacific; and survived all attempts by rival whaling nations to scuttle their fleet. All this restless activity came to an end, however, when the last of the bowhead whalers retired from the hunt. American ports fell idle. Ships that foundered in the ice were not replaced. Those that survived were sold and converted to other, more profitable uses. Elsewhere, whaling entered its modern phase, utilizing the machinery of industrialism to slaughter animals on an unprecedented scale. But Yankee seamen and merchants would not be part of this new age.

And what of Thomas Roys, the intrepid captain who opened up the last frontier of American whaling by daring to sail through the Bering Strait? He spent the last half of his life trying to perfect a weapon that would kill rorqual whales—the speedy, elusive species that could not be captured using traditional methods. Convinced, quite rightly, of the importance of rorquals to the future of whaling, Roys experimented with a harpoon propelled by a rocket from a shoulder launcher, instead of the more conventional style of harpoon cannon that resembled a heavy rifle. He dedicated all his time and resources to the rocket harpoon. On one occasion, in the Bay of Biscay, an experimental model exploded as Roys fired it. "Looking around me, I enquired who was hurt," he recalled. "There was no reply. I then saw lying upon the deck a finger with a ring upon it which I knew, and looking I saw my left hand was gone to the wrist. . . ."[19]

The loss of a hand did not stop Roys. Backed by a wealthy New York fireworks manufacturer, he established a shore-whaling station on the coast of Iceland where he continued to tinker with his rockets. But the weapon was unreliable, and Roys returned to the United States where he began a decade of wandering, still trying to work out the imperfections in his equipment. His final years are obscure. At the end of 1876 he washed up at Mazatlan, a Mexican fishing port, suffering from yellow fever. The old whaling captain died on January 29, 1877, indigent and unknown to anyone in the village. The American consul collected donations to pay for his burial. His estate amounted to less than two hundred dollars.[20]

PART TWO

WHALING IN THE MODERN AGE

INDUSTRIAL WHALERS

Now it is all over with the poor whales; the weapon cleaves them like fate, making an internal wound about 10 feet in diameter, closing at once every artery of life.[1]

–Captain Thomas Roys, 1860

During the decade of the 1860s, whaling entered the modern age and changed forever. The new machinery of the Industrial Revolution was applied to the chase with devastating results. Steam and explosives increased the speed and ferocity of the hunt. Species of whales that formerly had been ignored as unproductive or beyond the reach of the hand-held harpoon were now targeted for virtual extermination. New nations joined the chase, opening vast new hunting territories. Whale oil, no longer used for lighting, found a new market as an ingredient in soap and margarine. A boat steerer from a wooden whaler of the early nineteenth century would not have recognized what his vocation had become by the beginning of the twentieth.

For the romantic, the new world of whaling was a disappointment. It was all clanking machinery and cannon fire. There was the same hard labour, but little of the same risk. The odds were now weighted heavily in favour of the hunter. For all his skill, the gunner firing down on a whale from his perch in the bow of a catcher boat could not match the reckless courage of the harpooner riding on the very back of the great beast before driving his iron home. It was like sending a bullfighter into the ring armed with a cannon.

But whaling could not stand still to please the romantics. Right whales and grays hovered at the edge of extinction. In the Pacific the American fleet was running out of sperm whales, while in the Arctic the bowhead were fast disappearing. Old-fashioned whaling by boat and harpoon had come about as far as it could.

Whalers recognized that if their industry was going to survive they were going to have to find the means of exploiting the vast herds of baleen whales that so far had not fallen victim to their harpoons. These were the rorquals–humpbacks, fin-backs, blues and sei whales–that the whalers had always ignored because they were faster than a whaleboat and usually sank to the bottom when killed. "These whales," reported Thomas Roys in 1854, "will not generally allow a boat to come nearer than three or four rods of them, hence the difficulty of fastening to them which prevents our getting them at the present time."[2]

Rorquals take their name from the Norwegian *røyrkval*, meaning, roughly, "the whale with pleats," a reference to the furrows that extend back along the animal's throat and belly. As the mouth of a feeding rorqual opens, these furrows expand like the pleats of an accordion to accommodate huge amounts of food and water.

The family of rorquals includes the largest of all the great whale species, the immense blue whale and its only slightly smaller relation, the fin. Yet for all their bulk–and they weigh in at anywhere from 70 to 120 tons–they are also the slimmest and swiftest of all whales. The blue, for instance, can reach speeds in excess of twenty-five miles per hour; fins and seis are even faster. For centuries, their elusive speed was enough to protect these animals from capture as they roamed the world's oceans, migrating from their calving grounds in the temperate waters of the middle latitudes to their feeding grounds in the cooler waters toward the poles.

But the rorquals were not allowed to enjoy immunity forever. By the 1860s, Thomas Roys and other innovators were busily applying the machinery of the industrial age to the hunt. When they finished, they had not only developed a technique for catching rorquals, they had transformed the nature of whaling itself.

174

Rocket harpoon. The harpoon developed by Thomas Roys was fired from a bazooka-like launcher balanced on the harpooner's shoulder. It never proved effective.

I

During the summer of 1863, shipbuilders at the Nylands Verksted shipyard near Oslo were putting the final touches on a sleek eighty-two-foot steam schooner. It was a whaling vessel, but one totally unlike the familiar sailing ships that had long dominated the industry. Christened the *Spes et Fides* ("Hope and Faith"), it weighed a mere eighty-six gross tons, but it was powered by a steam engine that drove it through the water at a speed of seven knots. Fast enough to stalk the whales itself, it came equipped with a harpoon cannon mounted on its bow. Instead of the usual wooden boards, its hull was made of steel plate. A powerful, speedy hunter, it was the first of the modern whale catchers.

The inspiration for this unusual vessel came from a Norwegian sea captain named Svend Foyn. Born in 1809, Foyn grew up in Tønsberg, a small harbour near the head of the Oslo Fjord. His father died at sea when the boy was just four years old and life was a struggle for the widow and children. Undeterred by the fate of his father, Svend decided to become a seaman himself, embarking on a career in the merchant navy. In 1846 he left the coastal freight business to inaugurate the Norwegian seal hunt near Jan Mayen Island north of Iceland.

On his trips to and from the sealing grounds, Foyn took note of the herds of fin and blue whales that migrated, unmolested, along the shores of northern Norway. When the profits from the seal hunt began to decline in the early 1860s, he decided to give whaling a try. He was fifty-five years old and sealing had made him rich, but he was ready for a new challenge, not retirement.

A deeply religious man, Foyn in many ways resembled the stern Quaker moralists who made Nantucket the whaling capital of the world in an earlier age. His faith drove him to action. For him, business enterprise had a spiritual purpose. "God had let the whale inhabit [these waters] for the benefit and blessing of mankind," he later wrote, "and consequently I considered it my vocation to promote these fisheries."[3]

Foyn's first innovation was the *Spes et Fides* itself. Arctic whalers regularly installed auxiliary steam engines in their sailing vessels, but once on the grounds they continued to pursue the hunt from traditional rowing boats. Foyn broke new ground by combining the functions of boat and ship in one steam-powered, highly manoeuvrable vessel, fast enough to chase down the whale by itself. These swift catcher boats roamed the coastal waters in search of their prey, then towed the carcasses back to shore stations for processing. In a sense, it was a throwback to the earliest days of whaling when Basque hunters in boats patrolled the Biscay shoreline. The difference was that while the Basques were content to capture one or two animals, fleets of steamers like the *Spes et Fides* were soon slaughtering several hundred whales in a season.

Foyn took his new vessel north to the Varanger Fjord, a large inlet off the Barents Sea in Finnmark, Norway's most northerly province. Here he began to experiment with the equipment that eventually won him fame as the virtual inventor of modern whaling. Success did not come easily. The concept was straightforward enough–he would catch the whales by shooting them with an exploding harpoon launched from the deck of the catcher boat, then tow them back to his shore station in the fjord. In practice, however, there were myriad problems to be solved.

Whalers had been attempting to develop an effective harpoon

First station. Svend Foyn's whaling station in the Varanger Fjord in northern Norway.

gun for more than a hundred years. As long as the gun was fired from a small boat tossing in the waves, its accuracy was strictly limited. Besides, enough of these primitive guns exploded prematurely to leave harpooners very wary of them. By the 1860s most of these drawbacks were overcome, and the darting gun and bomb-lance had worked their way into the whaler's armoury. It was Foyn's achievement to take this process one step further by mounting a cannon on a swivel in the bow of the chase vessel. This weapon fired a harpoon with a grenade screwed to the tip. The grenade exploded a few seconds after the harpoon penetrated the whale, killing the animal outright or at least wounding it.

It took Foyn several seasons to develop a harpoon that penetrated the whale deeply, held firmly and exploded when it was supposed to. At one point in his experiments he paid a visit to Iceland where Thomas Roys was experimenting with his rocket gun. This weapon fired a harpoon from a shoulder launcher, much like a bazooka. Foyn recognized that Roys was on the wrong track—"not efficient and too expensive" was his opinion of the rocket launchers—but he was impressed by the American's approach to two other problems.

Once attached to an animal, whale line had to be strong enough to absorb tremendous shocks as the whale fought for its life and the boat bucked and heaved in the waves. Roys had worked out a system of rubber shock absorbers for relieving some of the strain on the line. Foyn later adapted this "rubber rope" for his own use. As well, he installed a steam winch on the *Spes et Fides* for hauling sunken whales off the sea bottom.

By the summer of 1868 the basics of Foyn's system were in place and he began to meet with success at his shore station in Finnmark. That season he killed thirty whales; by the end of the next decade his annual catch had jumped to ninety animals. For several years Foyn enjoyed a virtual monopoly on whaling in Finnmark. Then his success attracted the attention of other mariners in his native Tønsberg and the nearby village of Sandefjord, destined to become the whaling centres of Norway. By 1883 there were sixteen whaling stations in operation on the Finnmark coast, processing whales brought in by twenty-three catcher boats.[4]

II

Active to the end, eighty-five-year-old Svend Foyn was sponsoring an expedition to Antarctica in search of whales when he died in 1894. The patriarch of Norwegian whaling ended his days a very wealthy man, able to leave four million kroner to his favourite religious missionaries. But his greatest legacy was the technique of whaling that he left to his country. Norway was now indisputably the world leader. The Americans, once so prominent, were preoccupied with the development of their western territories and had all but forsaken whaling in the wake of the Civil War, with the exception of the Beaufort Sea fishery. The British still sent a token fleet to the Arctic each summer, but it barely managed to pay for itself.

Meanwhile, Norwegian whaling spread like wildfire. As the number of rorquals began to decline off Finnmark, the Norwegians transplanted their operations across the North Atlantic. First they migrated to Iceland in the 1880s, then to the Faröe Islands and Newfoundland in the 1890s and the Shetland Islands and Spitsbergen after the turn of the century. Stations were also established around the coast of South America and up the Pacific coast of North America. In 1908 a feverish outburst of Norwegian-style whaling began on the southern coast of

Africa, first near Durban and then around the Cape at several harbours as far north as Gabon. Most of these African whales were humpbacks, and by 1913 they accounted for almost a third of all the whale oil produced in the world.[5]

In most cases it was not just Norwegian technique but also Norwegian capital and manpower that fuelled this worldwide expansion. Other countries simply provided the whales. With no investment tied up in old-style whaling, the Norwegians were free to embrace the new. A northern people, with centuries of experience as sailors and fishermen, they were no strangers to the perils of the sea. From the first the finest gunners were trained in Norway, along with most of the crews that manned the catchers. Tønsberg, Larvik and Sandefjord, a trio of port towns south of Oslo, orchestrated the boom. Just as the whaling harbours of New England provided the manpower that once crewed the world's whaling fleets, so the southern county of Vestfold overlooking the Christiana Fjord produced most of the crews and ships that went whaling in the modern period. "Modern whaling began as a Norwegian industry and expanded as a Norwegian industry," writes British whaling historian Gordon Jackson. "When the time came for it to spread abroad it still depended upon Norwegian personnel; and Norwegian was the universal language of whaling. . . ."[6]

Whether they huddled on the storm-lashed coast of Finnmark, nestled in a green inlet on the Shetland Islands, or baked on the hot sand beaches of South Africa, modern whaling stations came out of the same mould. Each was home base for a small corps of three or four catcher vessels, which increased in size, speed and range as the years passed. Built low in the stern and always awash with water in heavy seas, these boats rose to a high bow where the gunner stood behind the cannon looking down on his target. Their rounded bottoms made them easy to turn after a fleeing whale, but the vessels bounced around in the waves like a frightened colt. "There can be few motions in the world more upsetting than the whirling dervish dance of a whale catcher," reported one newcomer after his first ride. "She rolls and pitches at the same time and there seems to be a rotary component in her performance which imparts giddiness as well as nausea, and a profound longing for easeful death."[7]

Each catcher carried a crew of about a dozen men, the most

Cannon. The cannon fired a deadly harpoon that exploded inside the whale.

important of whom was the gunner. On his skill depended the success of the enterprise. He had to position the boat precisely where the whale was likely to surface after a dive, motioning to the wheelhouse when to increase speed, when to ease off. Then, bracing himself atop the bucking platform, his view obscured by seaspray and breaking waves, he had to fire the harpoon into the neck of the great beast, exposed at the surface for just a few seconds. If he aimed too high, the hundred pounds of steel and explosive bounced harmlessly off the animal's smooth back; if he aimed too far back, the harpoon might lodge in the tail where the grenade did little harm.

Ideally, the harpoon penetrated to the whale's vital organs so that when it exploded death was instantaneous. However, this seldom happened. Instead the whale, wounded but still strong, dove for safety and began to run. The Nantucket sleighride, of course, was a thing of the past. No whale could tow a hundred-ton catcher the way it could a flimsy cedar rowboat. Nonetheless, the animals still put up a gallant fight. The record for the most difficult catch must surely belong to a giant blue whale harpooned off the south coast of Newfoundland early one morning in 1903. This animal was so strong that it towed the

180

catcher boat along at a speed of six knots, even though the vessel's engine was running half-speed in reverse. By evening the crew had worked the line around so that it was attached to the stern and the boat was towing the whale full speed ahead. This kept up all night, the whale still strong enough to pull the stern of the boat almost underwater. The next morning, twenty-eight hours after taking the harpoon, the whale finally lay still on the surface and the hunters were able to get close enough to kill it.[8]

The whale line on the modern catcher ran beneath the gunner's platform where it passed through a set of springs and pulleys called the accumulator, a sophisticated version of Roys's "rubber rope." The accumulator absorbed much of the strain on the manila line so that it did not snap, much as a fishing rod bends at the pull of a fish. The line passed up through more pulleys on the boat's mast, then back down to a steam winch mounted on the deck. To continue the analogy with fishing, the winch acted as the reel. As the whale tried to escape, the winch applied a constant pull on the line, reeling in the rope whenever the exhausted animal paused in its flight. At the end, when the whale died and sank, the winch had the all-important job of hauling the massive carcass to the surface.

New-style whaling depended on a series of innovations, like the exploding harpoon head and the accumulator, each as necessary as the other. Another of these breakthroughs occurred in the 1880s when whalers began inflating their catch to keep it afloat. Air was pumped in through a hollow pipe inserted into the animal's abdominal cavity. This allowed carcasses to be flagged and left while the catchers pursued more whales. Later the boat returned to collect the swollen corpses and towed them back to the station.

Everyone who visited a modern whaling station remarked first on the smell. It was suffocating. It filled the nose and made the eyes water. Clothes and skin reeked of it. It was the stench of decay, of rotting flesh, of the half-digested contents of whales' stomachs, of fetid pools of blood. "Flesh and guts lay about like small hillocks and blood flowed in rivers," reported one dismayed visitor.[9] But it was also the smell of money, and people quickly got used to it.

When a whale dies its body grows warmer, not colder, and its innards begin to decompose. The insulating layer of blubber keeps gases from escaping and raises the internal temperature of the carcass rapidly, speeding the process of decomposition. As the corpse decays, the quality of its oil deteriorates, so it was imperative to process the animal as soon after the kill as possible.

Whales generally arrived at the station during the night, allowing catchers to return to the hunt by daybreak. Very early in the morning the first bloated carcass slithered up the ramp onto the station's "plan" and the flensers set to work. The "plan" was a large, open space surrounded on three sides by boilers, winches and conveyers. Covered with a board floor and swimming in blood and entrails, it formed a huge butcher's block where the flensers removed the coat of blubber. Wielding sharp knives, they made several longitudinal slices down the length of the whale. Metal cables were attached and winches slowly drew off the blubber strips like the skin of a banana. As the blubber peeled back, flensers cut away the white connective tissue that bound it to the meat beneath. Eventually all that was left was a torpedo-shaped mass of bloody meat, viscera and bones covered in shreds of white tissue, much as the pith of a peeled orange covers the fruit.

Now it was the turn of blubber cutters who sliced the long strips of greasy blubber into rectangular blocks that were then fed into a revolving slicer that reduced them to small pieces. A conveyer carried the pieces to the cookers. Old-fashioned trypots had been replaced by open steam boilers in about 1870. The new technology exposed the blubber to hot steam, extracting the oil, which then drained away into purifying tanks.

In the early days of modern whaling, blubber alone was processed. The rest of the carcass was thrown away. Yet all parts of the whale are impregnated with oil. In fact, the meat and bones contain anywhere from 40 percent to 60 percent of the animal's total oil content. In other words, whalers were throwing away as much as one-half of the value of each animal.

Gradually the industry became less wasteful and stations were equipped to process every bit of the carcass. Once the blubber was removed, the remains were rolled across the "plan" and set upon by a party of "lemmers" who cut off the head, removed the baleen, severed ribs and backbone, drew out the viscera and butchered the meat.

This was not a pleasant job. By now the corpse swelled with foul gases. When the stomach broke open it spilled a vast cataract of undigested food around the legs of the "lemmers."

And this was not the worst of it. The intestines were like giant worms, bulging with the pressure of decomposing fecal matter. "One touch of the knife, purposeful or accidental, on these protuberances," explained F. D. Ommanney, a scientist at one of the shore stations, "and an explosive jet, a fountain or cascade, . . . shot out throwing orange liquid intestinal contents six feet and more from the carcass."[10]

Meat, bones and viscera all went into cookers. Once their oil was removed, the residue was pulverized to make fertilizer. Meanwhile, baleen was cleaned, dried and bundled for shipping. The model of efficient organization, a modern whaling station could process twenty carcasses in a day's work. The "Norwegian method" reduced production costs low enough that whale oil again became competitive with other oils on the market.

III

While Norwegians spread their new style of industrial whaling around the world, their own industry on the shores of Finnmark quickly foundered. In part it was the familiar story of overhunting. But more important was the opposition of the fishermen of Finnmark. They were established on the coast long before the whalers came, harvesting the inshore stocks of cod, capelin and other fish. Soon after the first whaling stations opened, the fishermen complained that their catch was falling off. They blamed the whalers who, they said, were killing off the whales that drove the fish inshore from the deeper offshore waters.

Although there was in fact no proof that whales behaved like shepherds, herding their flocks of fish landward, the fishermen won the ear of the government. In 1880 legislation stopped Svend Foyn's catcher boats from operating between January 1 and May 31 in the Varanger Fjord and within a mile of the northern coast. Unsatisfied with a partial ban, the fishermen kept up their agitation. Early in the new century their cause seemed to be vindicated as the Finnmark cod fishery failed miserably. In June and July 1904, angry residents took to the streets of the small fishing villages and during several nights of rioting completely wrecked one of the whaling stations. Troops restored order and an alarmed Norwegian parliament banned all whaling off Finnmark for ten years.

Elsewhere local fishermen and whalers coexisted uneasily as well. In Iceland, the Danish government, which then controlled the island, ruled that all whalers had to be naturalized citizens; then in 1913 the hunt was banned entirely. And in the Shetland Islands and the Hebrides the complaints of herring fishermen succeeded in winning restrictions on the activities of whalers.

Ultimately it was not the antagonism of fishermen that deflated the boom in shore whaling at the outbreak of the First World War. During the explosion of new-style whaling that followed Foyn's innovations, the industry simply over-expanded. To take just one example, in Newfoundland, where a Norwegian company took its first whale in the summer of 1898, there were fourteen stations operating by 1904, along with another thirteen along the coast of Labrador. Not surprisingly, the returns at each station plummeted, and one after the other they closed, until by 1915 only three remained in business.[11]

The pattern repeated itself all around the North Atlantic and on the coast of Africa as well. But as usual, when one whaling ground failed, another was discovered to take its place. As the new century dawned, whalers were looking to yet another new frontier where vast numbers of the giant rorquals congregated each year: the Antarctic, the last, best, killing ground.

WHALING IN THE SOUTHERN OCEAN

Adventure, that is whaling in the waters around the Antarctic continent, the greatest and most wonderful adventure of our age. . . . Down here in the very shadow of the South Pole a livelihood has been created which every year directly and indirectly earns hundreds of millions of kroner. The Norwegians have made the way through the hundred-mile wide belt of ice into the Ross Sea a thoroughfare. They have taken a gamble and they have won. Whaling in the Antarctic Ocean is undoubtedly the toughest and most hazardous livelihood in the world.[1]

–Carl Ben Eielson, Norwegian-
American aviator, 1929

The island of South Georgia looms out of the sea fog like some wild, ice-encrusted range of mountain peaks. Located on the desolate, southern rim of the Atlantic, 1,275 miles east of the tip of South America, it provided a home for cumbersome elephant seals, stately penguins and flocks of nesting seabirds, but not for human beings–at least not for many years. Captain James Cook and his crew aboard the *Resolution* were the first outsiders even to set eyes on the island. It was late in Cook's second great voyage in search of a southern continent. He had already sailed closer to the South Pole than anyone before him, but the ice drove him back and he took consolation in surveying the desolate islands that guard the entrances to Antarctica.

As he came within sight of South Georgia in January 1775, Cook for a moment wondered if perhaps he had found the

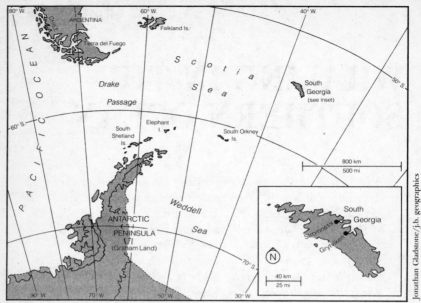

Antarctic Grounds. Shore stations at South Georgia.

continent he sought. The coast, which rose steeply in high cliffs from the water's edge, was broken by deep, jagged fjords clogged with tongues of ice that flowed like rivers from the inland glaciers. The interior was a jumble of towering mountains, some rising 9,000 feet. If this forbidding landscape was Antarctica, Cook wryly observed, then the continent "would not be worth the discovery."

But as the *Resolution* sailed along the coastline, South Georgia–Cook named it for King George III–revealed itself to be an island, about one hundred miles long, shaped vaguely like a crescent. The expedition landed on the north coast. "The wild rocks raised their lofty summits till they were lost in the Clouds," wrote Cook, "and the Vallies laid buried in everlasting Snow. Not a tree or shrub was to be seen, no not even big enough to make a tooth-pick."[2] Cook went through the motions of claiming the island for Britain, making it the first British territory in this remote, southern region. But members of the expedition could not have been sad to leave South Georgia. "The very thought to live here a year fills the whole Soul with horror and despair," concluded the naturalist Johann Reinhold Forster.[3] "It is thirty-one leagues in length, ten in breadth and of less value than the

186

smallest farmstead in England," added his colleague, Anders Sparrman.[4]

Despite South Georgia's isolation and severe climate, however, Anders Sparrman turned out to be wrong. The island soon began to attract a steady stream of visitors, chiefly seal hunters who came to harvest the fur and elephant seals that lounged along the beaches in vast numbers. Then, more than a century after Cook's arrival, northern whalers appeared. In the Arctic the stocks of bowhead were fast disappearing, while Norwegian hunters in the North Atlantic were noticing that rorquals were becoming more elusive. The industry badly needed a new source of high-quality oil.

Still not convinced that they could prosper on rorquals alone, European whalers were looking for an undiscovered herd of right whales, one that they had not already decimated by over-hunting. They recalled that Sir James Clark Ross, on his voyage to the Antarctic in 1841, reported seeing large numbers of right whales in the Ross Sea. "Wherever you turned your eyes," wrote the explorer, "their blasts were to be seen."[5] But Ross was wrong. These were not right whales that he saw, and the first whalers who came south were in for a big disappointment.

In 1892 a total of five whaling vessels set off to investigate the southern ocean. Four were Scottish; the fifth was a Norwegian ship commanded by Carl Anton Larsen, an experienced Arctic whaler and sealer. The small flotilla combed the Weddell Sea in search of right whales, with no success. After taking a few seals, the ships returned home to make their report and count their losses. Captain Larsen returned to the Antarctic the next season, with equally discouraging results. There was no hidden source of right whales waiting to be exploited.

Yet Larsen had seen enough to convince him that the Antarctic did represent the future of the world whaling industry. Right whales might be rare, but rorquals were plentiful. The region had a short hunting season. It was remote from the whaling ports of Norway, and navigation in the ice was perilous. But Larsen believed that if he could find a suitable harbour for a shore station these obstacles could be overcome and modern, Norwegian-style whaling might transform the Antarctic into the next great whaling frontier.

Grytviken. The loneliest whaling outpost in the world, Grytviken is situated on the forbidding coast of the island of South Georgia.

I

Whalers arrived in Antarctica just as a great age of exploration was getting under way there. The polar treks of Scott and Amundsen were still in the future, but during the 1890s the first explorers landed on the Antarctic mainland and began charting its intricate shoreline. Among the swarm of adventurers who took up the challenge of discovery in the Antarctic was a Swedish geologist, Otto Nordenskjöld. Nordenskjöld theorized that at one time South America and Antarctica were joined. In 1901 he set out for Graham Land, the long peninsula that reaches out like a crooked finger from the southern continent toward Cape Horn, to prove his theory. As captain of his ship, the whaler *Antarctic*, he wisely chose Carl Anton Larsen.

After failing to penetrate the ice of the Weddell Sea, Larsen dropped Nordenskjöld and his party on Graham Land and proceeded via the Falkland Islands to South Georgia where he spent two months carrying out a program of scientific research. Larsen visited Cumberland Bay on the island's north shore and discovered a snug little harbour formerly used by the sealers. He named the spot Grytviken ("Cauldron Bay"), after the boiling-pots abandoned by the sealers, and kept it in mind for the whaling venture that he was beginning to plan.

First Larsen had to retrieve Nordenskjöld. But on its way to Graham Land for the rendezvous, the *Antarctic* fell victim to the ice and sank. Larsen and his crew escaped to a small island where they survived on seal and penguin before finally trekking across the ice to Nordenskjöld's camp. An Argentinian gunboat rescued the entire expedition and took it to Buenos Aires where the explorers received a heroes' welcome.

During his stay in Buenos Aires, Larsen promoted his idea for whale hunting in the southern ocean. He managed to convince three foreign residents of the city to put up enough capital to form the Compania Argentina de Pesca. Appointed manager of the company, Larsen returned to Sandefjord to outfit an expedition to South Georgia. By November 1904 he was back at Grytviken with sixty men, two sailing ships, a steam catcher boat and building materials and equipment to erect a station. On December 22 the newcomers killed their first whale, and two days later the first oil produced in Antarctica drained from the huge cookers. During its first year in operation, Larsen's company took 183 whales. The animals were so plentiful

Butcher shop. A blue whale lies on the flensing deck at Grytviken during the 1925 season.

that production was limited only by the capacity of the station to process them.

Word of Larsen's success spread quickly back to Europe. Other companies arrived to set up factories. The British government, which claimed to own South Georgia, recognized that it would have to control development on the island and began issuing licences to the whalers. These permits allowed companies to establish a station in one of the harbours and to operate a limited number of catcher boats. Some operators did not even bother to build a shore station, preferring instead simply to park a large cargo vessel in the harbour. These floating factories, usually converted merchant ships, contained several blubber cookers. Catchers brought in their whales, which were then flensed and processed much as they had been in the days of old-style whaling. In the years before the First World War, floating factories grew in size until they weighed several thousand tons.

The British were only willing to issue licences for seven stations on South Georgia. In 1909 these concessions were all taken, leaving no more room on the island. Latecomers migrated to the South Shetland Islands, then to the South Orkneys. By the 1912–13 season there were sixty-two catchers at work in Antarctic waters, killing almost four times as many whales as hunters still active on the traditional northern grounds. In that season alone, 10,760 whales went through the cookers at the southern harbours.[6]

The early Antarctic hunt was extremely wasteful. Flensers removed only the thickest, most oil-rich layers of blubber. The rest of the carcass was cast adrift to wash up on the shore where it slowly rotted. When F. D. Ommanney visited South Georgia in 1929, he described beaches "edged with ramparts of bleached bones, skulls, jaws, backbones and ribs, memorials to that uncontrolled slaughter."[7] Recognizing the appalling waste, British officials attempted to put a stop to it by requiring as part of the licence that whalers process all parts of the carcass. Within a few years of Larsen's arrival, the worst excesses were over.

II

With the arrival of the whalers, South Georgia became the most southerly inhabited spot in the world. A less hospitable environment would be hard to imagine. The island is the summit of a submerged mountain range. Lofty peaks descend steeply to the surf-beaten shore. Mount Paget, the highest peak, towers 9,625 feet above sea level. Ice and snow cover more than half of the island permanently. On almost every day of the year, some rain or snow falls. Only the sparsest vegetation survives the cold. There are no trees. The largest growing things are clumps of tussock grass sprouting close to the water. In summertime, green lichen brightens the rocky slopes. The whaling stations were located on the north side of the island, where fierce gusts of wind howling down the mountainsides routinely threatened to blow the buildings from their narrow beaches. The mountains loomed up directly behind them and more than one whaler lost his life when an avalanche cascaded down onto the station.

"Although the island is a British possession," remarked one visitor, "one might imagine oneself in Norway."[8] Most of the seven hundred or so residents were indeed Norwegian—sailors, flensers, gunners, lemmers and labourers who migrated from their homes in the county of Vestfold each November for a brief southern summer of hunting. The work was hard. Catchers stayed at sea almost constantly, their crews living in cramped forecastles that tipped and swayed in the rough southern waters. At the stations, crews began their workday when the whistle blew at six o'clock in the morning and did not knock off until it blew again twelve hours later.

The whaling companies did not make much allowance for creature comforts. Living conditions were spartan and unclean. R. B. Robertson, who came as a doctor in 1950, described South Georgia as "the most sordid, unsanitary habitation of white men to be found the whole world over, and the most nauseating example of what commercial greed can do at the expense of human dignity."[9] There is no reason to suppose that accommodations in the early period were any better.

The remote stations offered little for the men to do in their off hours. There were no libraries or recreation facilities. Mountain climbing was dangerous; the only game for hunting were seals and penguins. The importation of liquor was supposed to be regulated but homemade stills were common. One particularly lethal potion was concocted out of shoe polish.

Outside visitors provided a welcome distraction. Concerned about the future of the whaling industry, the British began a program of scientific investigation in and around South Georgia in the 1920s. Called the "Discovery Investigations," after the ship that conducted many of the experiments, the program lodged its scientists at a permanent station next door to Grytviken.

South Georgia was also a jumping-off point for polar explorers on their way to and from the icy continent. Notable among these adventurers was the British explorer, Ernest Shackleton, who attained legendary stature among the whalers on the island. Shackleton arrived at Grytviken in November 1914 on his way to attempt the first crossing of Antarctica. While he waited for the last mail from Europe to catch up with him, the explorer fell under the spell of the rough glamour of whaling and began to make plans to form his own company when he returned from his expedition.

After leaving South Georgia, Shackleton and his men sailed south through the Weddell Sea in their ship *Endurance*. Before they could reach the mainland, however, the vessel was beset in the ice and carried away from land. For eight months the expedition drifted with the current. Finally the *Endurance* was overwhelmed and sank. The crew made their way north on foot to the ice edge, then by boat to Elephant Island off the tip of Graham Land. It was the first land they had set foot on in more than a year. But their ordeal was far from over. Uninhabited, barren, remote, Elephant Island was no place to spend a winter. Shackleton decided to try to reach South Georgia, more than seven hundred miles to the northeast. In a twenty-two-foot open boat, the explorer and five of his comrades set sail. Seventeen days later, exhausted, starving and almost frozen to death, they arrived on the deserted south shore of South Georgia. Unable to sail the boat any farther, the expedition landed and set up camp.

No one had ever crossed the island before, but Shackleton realized he would have to make the attempt. Leaving tents, sleeping bags and almost all their gear behind, he and two others gambled everything on a lightning dash for safety. Clambering over high mountain passes, sliding down the sheer face of glaciers, slogging across open fields of deep snow, the trio trekked for thirty-six hours virtually non-stop. When they stumbled into the Stromness whaling station late the afternoon of the second day, caked with grime and gaunt with hunger and fatigue, the whalers who had seen them off on their great adventure seventeen months earlier did not even recognize them.

Shackleton felt a kinship with the men of the South Georgia

whaling stations and took a special pleasure in sharing the story of his misadventures with them. "Under Providence we had overcome great difficulties and dangers," he wrote, "and it was pleasant to tell the tale to men who knew those sullen and treacherous southern seas."[10] For their part, the whalers returned this affection. In years to come they loved to regale newcomers to the island with the saga of Shackleton's dramatic escape. Early in 1922, the explorer was back on South Georgia preparing for yet another expedition when he suddenly died. He was buried at Grytviken and successive generations of whalers took pride in tending the modest grave.

III

The Antarctic whaling industry was based on a reddish-pink, bug-eyed creature that looks like a shrimp and grows no larger than a human thumb. Known popularly as krill, these tiny crustaceans swarm on the surface of the southern ocean in huge numbers, staining the water reddish-brown in vast carpets of squirming life that extend several miles in every direction.

Krill thrive only in the cold, oxygen-rich waters south of the Antarctic Convergence, the dividing line between polar waters and warmer currents from the north. They appear quite suddenly with the arrival of summer, feeding on the microscopic plant life that is nourished by the lengthening days of sunlight. During this seasonable smorgasbord, lasting only until the return of winter, the giant whales gorge themselves, laying on the thick layers of blubber that will sustain them through their breeding season, when food is relatively scarce. Lolling through a cloud of krill with its mouth open, a blue whale might eat as much as eight tons of the protein- and vitamin-rich shrimp every day, putting on weight at a rate of a ton every ten days. It has been estimated that before their numbers were reduced by hunters, southern whales devoured two hundred million tons of krill a year.[11]

Located as they were at the heart of the "whales' larder," the whalers of South Georgia had only to wait for their prey to appear each November. During the first decade of the Antarctic hunt, the catch consisted overwhelmingly of humpbacks. A slow swimmer that tended to pass close to shore, the humpback fell easy victim to the whaler's harpoon. By 1918 they were virtually wiped out. Belatedly, the British banned the killing of humpbacks.

But hunters were already turning their attention to another species, the blue whale.

Whaling reached its modern peak at the expense of the blue. About 150,000 of these giants visited the Antarctic annually in the years before the hunt began. They were prized by the whalers because they were so big. Southern blue whales were killed that weighed over 150 tons, reaching lengths of more than ninety feet. Today, estimates put their population worldwide at a mere 10,000 animals; one alarming survey has concluded that only a few hundred blues survive in the southern ocean.[12]

The slaughter was relentless. Between the two world wars, more than 20,000 blues died each year. In one season, the southern summer of 1930–31, the kill reached a record 29,400 animals.[13] It was the blue whale that became the standard for the industry; all other species were measured against it. And it was the slaughter of the blue that prompted whaling nations to take the first faltering steps towards regulating the hunt and conserving the dwindling stocks of whales left in the world's oceans.

IV

The slaughter of the blue whale in the Antarctic coincided with the emergence of the pelagic, or deep water, factory ship, the technological development that spelled final disaster for the world's whales. The first floating factory based at South Georgia was the *Admiralen*, weighing 1,517 gross tons and needing a crew of sixty-three. By the 1930s, factories of 13,000 tons were common, with crews ten times as large. But it was more than its gigantic size that characterized the modern factory. The main difference between it and the early floating factories moored in their island harbours was that the pelagic vessels worked far out at sea, away from the restrictions imposed by licences and regulations. Whalers could follow their prey right to the ice edge, killing, flensing and processing their catch hundreds of miles from the nearest land.

At each turning-point in the early history of Antarctic whaling, the figure of Carl Anton Larsen appears. His ambition took him into every corner of the whaling islands. He probably knew more about the stormy southern ocean than any other living person. He is one of those characters, like Thomas Roys or Svend Foyn or William Penny, who never attained the fame of the great explorers, yet who, through perseverance and foresight, laid the foundations for the economic exploitation of the world's frontier regions.

Larsen, after managing the station at Grytviken for a decade, resigned from the company he had founded and returned to Norway in 1914. But a career that began in the seal rookeries of the North Atlantic had not yet run its course. Larsen wanted to return to the Antarctic and to whaling. In the familiar area of South Georgia, where the British had already handed out all their concessions, his way was barred. So he shifted his attention to the other side of the continent, to the Ross Sea, as yet untapped by whalers.

In 1893 Svend Foyn had sent a ship to the Ross Sea to look for whales. After many misadventures, the tiny vessel broke through the ice and succeeded in landing the first party to set foot on the Antarctic mainland. But the expedition found no whales and returned to Norway with an empty hold, a financial disaster. Undiscouraged by this example, Captain Larsen approached the British authorities to obtain a licence to go pelagic whaling in the Ross Sea. Not everyone acknowledged that Britain had sovereignty in the area and the issue dragged on. Finally Larsen received a concession, and in October 1923 he left Norway in the *Sir James Clark Ross*, a newly converted 8,223-ton factory with a fleet of five catcher boats and a crew of 130 men.

Pausing briefly in Tasmania, the *Ross* embarked for Antarctic waters at the end of November. As the ship ploughed southward, the crew prepared for the whaling in the time-honoured way–rigging gear, oiling winches, sharpening flensing spades, readying all the equipment for the appearance of the whales. Crossing the Antarctic Convergence, the temperature dropped suddenly and icebergs began to litter the ocean. "We stood upon the forecastle long into the night, drinking in with eager eyes the beauteous sights, the like of which we had never seen before," wrote A. J. Villiers, an Australian journalist who joined the expedition in Tasmania.[14] "There were some that seemed but great arches of ice built around caves in which every pure and beautiful shade of white and blue shimmered and danced, waxed and waned; there were bergs with long, low, dangerous, submerged ice-feet, across which the blue waters gurgled and foamed as they lifted on the gentle swell. There were wonderful, great floating masses of ice resembling, to an imaginative mind, cruel cold old bastions, feudal castles towered

CHAPTER 10

and battlemented in their pride, white marble halls, cathedrals;
some afar off floated ethereally like pure white clouds."[15]

On December 13, the *Ross* and its entourage of catchers
entered the belt of pack ice that rims the outer margins of the
Ross Sea. For a week the giant mother vessel rammed its way
through the floes, towing the string of smaller catchers behind.
"At first the floes of ice were not so large," recalled Villiers,
"–about the size of big pavement flags and as perfectly flat, but
as we progressed southward they grew and grew until each was
the size of a long irregular piece of a city's widest street, their
sides fitting closely one into the other and cemented together by
loose brash ice. Still they steadily grew until they reached the size
of gardens and parks, until finally they became one great solid
flat surface bathed in a glaring white light most trying to the
eyes, stretching away on every side to the distant horizon."[16]

Halfway through this obstacle of ice, the *Ross* became the
largest ship ever to have crossed the Antarctic Circle. As it did
so, the floes began to jumble up in massive pressure ridges where
wind and current had driven them together. The scene reminded
Villiers of heaps of sugar cubes "spilled carelessly on a spotless
white table-cloth."[17] In this heavier going the *Ross* became mired
for the first time, but the men used saws to cut away the ice and
the ship forced a path through. At last the flotilla broke out of
the pack and emerged into the open water of the Ross Sea.

On Christmas Eve, after three days of steaming south through
uninterrupted daylight, Captain Larsen's flotilla arrived at the
Great Ice Barrier, a high, white wall of ice that marked the end
of navigation. Eighty years before, James Clark Ross had gazed
upon the Barrier for the first time in history. It was a "mighty
and wonderful object," he exclaimed, "far beyond anything we
could have thought or conceived."[18] Rising up out of the ocean
more than 150 feet, the Barrier forms the edge of a vast shelf of
ice, fed by inland glaciers, that spreads out from the coast of
Antarctica, covering an area of the Ross Sea almost as large as
the Iberian Peninsula. Surf pounds against the base of the cliff,
raising a sullen and constant roar. The face of the ice is a
shimmering blue curtain, pitted with caverns and grottoes. Huge
pieces of the Barrier break off in bergs as high as ten-storey
buildings several miles across.

In places, sailors perched in the rigging could see over the
top of the barrier into another, alien world. The surface of the

Baleen drying. Strips of baleen were cleaned and dried before being bundled for shipment home.

ice shelf spread away toward the Pole, in the words of James Ross, "like an immense plain of frosted silver." Across this frozen, windswept table of ice, Roald Amundsen had reached the South Pole and returned, and Robert Scott had died in the attempt.

The Ice Barrier stretches almost five hundred miles across the Ross Sea from Victoria Land on one side to King Edward VII Land on the other. When the whalers arrived late in 1923 Captain Larsen recognized the truth of James Ross's warning that ". . . we might with equal chance of success try to sail through the Cliffs of Dover, as penetrate such a mass." The ungainly factory ship turned aside and began to trail along the Barrier looking for a sheltering harbour. On December 28 one of the catchers killed the first whale, a blue, and three days later the *Ross* steamed into Discovery Inlet, an opening in the Barrier that would be home to the expedition for the next two months.

While the mother ship stayed at Discovery Inlet, the five catchers roamed out beyond the barrier, killing whales and bringing them back to the factory for flensing. Most of the catch turned out to be large blue whales instead of the smaller humpbacks Larsen expected to find. Because of their bulk, blues could not be lifted onto the factory. Instead they were flensed in the water alongside, a dreadful job in the numbing cold.

Flensers worked from a small dory, or on the back of the carcass itself. Since they could not manipulate their cutting spades wearing gloves, they worked barehanded, using sawdust to improve their grip. Frostbite was common, even though they tried to warm their hands by washing them in the blood of the corpse. Perhaps the worst job belonged to the young boys who stood stoically in the dory for hours on end, holding it against the whale while the flensers carved up the blubber. Anyone tumbling into the water would be frozen stiff within minutes.

"Discovery Inlet is a very bad harbour," remarked Villiers matter-of-factly.[19] For one thing, it was a deep-freeze. Temperatures plunged as low as minus fifty-five degrees Fahrenheit; the whalers considered minus ten degrees a spell of balmy weather. "There were but two kinds of days," wrote Villiers, "bad days and worse. . . ."[20] Terrible gales from the south howled across the ice shelf, bringing blizzards of driving snow that piled up in drifts on the deck of the ship. "All ropes, chains and wires on deck became frozen beneath feet of ice, so that if they were required for use it was necessary to break them out with sledge-hammers and cold chisels. The rigging was festooned with frost and ice; icicles inches long hung from the wireless aerials."[21] On calm days, icefogs and mists enveloped the ship. The cold was so intense that ears and noses bled. Unlike the old-style wooden whalers with their rope rigging, the *Ross* was made of steel. "In those dreadful temperatures," Villiers pointed out, "to touch any of this steel with the bare hand often meant to lose the skin tearing one's hand free again; to lean against or to come into contact with anything meant to freeze fast to it."[22]

Despite these terrible conditions, whaling continued until March 7 when Larsen recognized it was time to leave for home. The whales had left the area and ice was gathering outside the harbour. After enduring a rough passage north, the whalers arrived safely in New Zealand a month later. Financially, the voyage was not a great success, though it did turn a modest profit. Whales had turned out to be less plentiful than Larsen expected, and for the most part too large to be lifted on board the factory for processing. And the weather was awful. The whole venture, concluded Villiers, was "one of infinite difficulty and danger."

Gunner. The gunner stood exposed to the elements on his platform in the bow of the catcher boat as it plunged through the stormy seas.

However, these problems did not stop Larsen from trying again. At the end of 1924 the *Ross* was back at the Ice Barrier, this time making enormous catches. The weather was mild, the whales plentiful, and the expedition returned with 31,500 barrels of oil.[23] On some days, so many carcasses were waiting to be processed that the catchers had to stop hunting to let the flensers catch up. Instead of tying up in a harbour, the factory cruised among the floes, thus inaugurating true pelagic ice catching. The only negative aspect of all this success was that Captain Larsen did not live to see it. He died of heart disease in his cabin aboard the factory shortly after the season began.

V

The pioneering voyages of the *Sir James Clark Ross* in the Ross Sea highlighted a lingering inefficiency plaguing the deep-water factories. The largest whales still had to be flensed alongside the vessel. Winches could not lift the massive, hundred-ton weight of a blue whale out of the water. Furthermore, when carcasses were small enough to lift on board, crews had a difficult time securing them so that they did not roll around the deck in heavy seas. Poor weather generally put a stop to flensing. But it was imperative to cook the blubber as soon after a kill as possible to obtain the finest oil. Any delay meant a deterioration in the quality of the product.

The problem was solved by a whale gunner from Sandefjord named Petter Sørrle. Early in the season of 1912–13, Sørrle was working with a floating factory near the South Orkney Islands. For several weeks the factory could not get in to the shore because of heavy ice. While large numbers of whales were visible from the deck, nothing could be done about killing them on the high seas. It was during those frustrating weeks, Sørrle said, that he came up with the idea of inserting a slipway in the stern of the factory so that carcasses could be hauled on board for processing. Sørrle spent the next eight years working at one of the South Georgia shore stations, but in the early 1920s he was able to patent his idea and in 1925 the floating factory *Lancing* arrived in the southern ocean, the first factory equipped with a stern slipway.[24]

The introduction of the slipway sparked a rapid spread of deep-water whaling by factory ships. By 1931, there were forty-one factories operating in the Antarctic, fed by a fleet of

205 catcher boats.[25] As the name implies, a factory took the raw material delivered by the catchers and transformed it quickly and efficiently into products for market. Whales were manoeuvred to the opening at the stern of the vessel where a huge steel claw took hold of the tail and hauled the carcass up the sloping ramp onto the flensing deck. Immediately, a team of flensers wielding curved blades the size of hockey sticks fell on the animal, carving off the blubber in long strips. The massive corpse was turned using steel cables manipulated by power winches.

The strips of blubber were sliced up into smaller lengths, then dumped through holes that led down to the boilers on the deck below. After flensing, the carcass moved ahead to the lemming deck where it was butchered. Meat and bones were cut up and sent below to the cookers. The liver was removed, chopped up and sent to the "liver plant" where it was treated to extract vitamins.

Crews on a factory ship worked around the clock during the hunting season. At peak efficiency, it took them less than an hour to dispose of a fifty-ton whale carcass.

As technology developed, less and less of the animal was wasted. The sticky residue from the cookers, called "glue water" or grax, was centrifuged to extract the last particles of oil. The remains of the meat and bones were dried and crushed to make meal for animal feed and fertilizer. Only the baleen, now quite worthless on the world market, and the entrails ended up back in the water.

VI

While they worked to increase the speed and thoroughness with which carcasses were processed, whalers also dreamed of the perfect kill, one that was instantaneous and left the corpse almost undamaged. The grenade harpoon was a grimly fatal weapon, but the explosion inevitably destroyed part of the meat and accelerated the process of internal decay that spoiled the oil. The search for an alternative led to some far-fetched experiments. Depth charges, nerve gas, and injections of compressed air, anaesthetics or exotic poisons were all proposed. But in the end, these methods always turned out to be impractical.

A particularly fantastical invention was the exploding submarine designed by Norwegian engineer Jens Andreas Morch. Morch was a well-respected whaling expert, but in this case his enthusiasm got the better of him. He proposed to launch an unmanned electric submersible, shaped like a whale so as not to alarm its prey. The noiseless, underwater vessel would approach and ram the swimming whale, then explode. The idea got no further than Morch's drawing board.[26]

One technique that has come closer to replacing conventional harpoon whaling than any other is electrocution. A jolt of electrical current would kill, or at least paralyze, the animal almost immediately without mangling its insides. An added advantage was that electrocuted whales tended to float, buoyed up by the air caught in their lungs when their muscles became paralyzed.

Experiments began in the 1850s when two German scientists patented an "electric whaling apparatus" that established the basic principles of electric killing. After the harpoon planted an electrical cable in the animal, the current went out along the cable, through the whale and completed the circuit by returning to the ship via the salt water. The German apparatus did not work, but over the years several attempts were made to perfect the system. During the 1930s Norwegian whalers in the Antarctic had a prototype in use that killed about two thousand whales before the technique was abandoned as too expensive. After the Second World War electrocution won the support of the animal welfare lobby, which believed it was a more humane method of killing than explosives. However, once again experiments failed to perfect a weapon that was both safe and effective.[27]

VII

Antarctic whaling was based on two of the most common household products—soap and margarine. With the introduction of petroleum in the nineteenth century, whale oil seemed to have a dim future. No longer in demand as a lighting fuel, it still found various uses in industry, but manufacturers disliked its fishy odour and disagreeable taste and chose other animal fats and vegetable oils wherever possible. In 1900 the world whaling industry, despite the advances of Svend Foyn, sold only 87,300 barrels of oil, about one quarter of what Yankee whalers alone were bringing home fifty years earlier.[28]

Then, in 1906, the world faced a sudden shortage in the supply of animal fats. Production of soap and margarine, two major users of fats and vegetable oils, outstripped the available supply of raw material. Forced to come up with alternatives, manufacturers took a new look at whale oil.

Soap producers were already using raw whale oil in small quantities in their low-grade, industrial cleansers where it mattered less when the oil turned black or rancid. However, manufacturers really needed a hardened fat, not a fluid oil. The chemical composition of the two is quite similar. The main difference is that in oil the fatty acids do not contain hydrogen atoms and are said to be unsaturated, while in fat the acids are saturated with hydrogen. The challenge for the whaling industry was a scientific one: how to saturate the oil with hydrogen and turn it into fat.

This bit of alchemy was perfected in the decade before the First World War. Through a process called hydrogenation, in which hydrogen was added to the unsaturated acids, whale oil turned into a whitish solid and lost much of its disagreeable smell, taste and colour. Soap makers could now use hardened whale oil in their better-quality products and margarine makers began mixing it with other animal fats. As well, a by-product of the process, glycerine, played an important role in the manufacture of explosives.

Consumers resisted the idea of eating and washing themselves in whale oil. "To look kindly upon blubber as a source of imitation butter," writes business historian Charles Wilson, "called for a more lively appreciation of the wonders of science than the average housewife could muster."[29] As a result, manufacturers did not broadcast what they were doing.

Consumers also resisted attempts to interest them in another whale product, the meat. Though it is rich in nutrients and tastes not unlike beef, whale meat never caught on with diners, who complained of its fishy smell and its oily texture. During the First World War, as meat supplies dwindled, a widespread advertising campaign was launched in an attempt to make "sea beef" more acceptable, especially to North American shoppers. But as soon as the war ended, so did any market for whale meat. The exceptions were Japan and Norway, where canned and fresh meat continued to sell. Otherwise, whalers simply ground it up and sold it for animal fodder and pet food.

As an ingredient in other products, however, whale oil slowly gained acceptance. Hydrogenation methods improved, removing the last, lingering smells and tastes of the sea. In 1929 scientists succeeded in producing a hardened whale oil that

needed almost no other ingredients to make a perfectly acceptable margarine. During the Depression, consumption of margarine increased, and this product alone used more than 80
percent of the world production of whale oil. Once again
whalers could go about their business confident that the market-
place would buy up as much oil as they could produce.

VIII

The 1930s ended with an unprecedented slaughter of whales.
Never before had so many of the animals fallen victim to the
whaler's harpoon. During the heyday of the nineteenth century
fishery in the Pacific, five thousand sperm whales died each year.
Yet in the season of 1937–38, hunters killed almost 55,000
whales, setting a record for tonnage that has never been
equalled. Between 2.5 and 3 million tons of the animals were
killed every year.[30]
   They were harvested much like any other natural resource.
Little regard was paid to their conservation. Chased down by
implacable catcher boats, processed aboard giant factory ships as
long as aircraft carriers, they were treated no differently than
the timber and minerals that industrial society also required for
its comfort and sustenance. They were similar in this regard to
the herds of wild buffalo that once blackened the North
American plains. In the latter half of the nineteenth century
hunters turned on these shaggy animals and quickly exterminated them for their hides. Today the prairie bison exists only in
tiny, scattered herds, protected in special preserves.
   Did the fate of the American buffalo presage the dismal
future of the whale? Many whalers thought not. "These giant
whale argosies cannot continue–they would not be profitable–unless they are assured of full cargoes, or nearly full," a
sanguine A. J. Villiers glibly assured his readers in the 1931
edition of his book. "They cannot all get full cargoes unless there
are plenty of whales. . . . Therefore, as soon as the whales thin,
these ships will have to leave them alone."[31] The whole history of
whaling contradicted Villiers; nevertheless, his was the conventional wisdom. "The view held in all well-informed quarters,"
stated a report to the giant soap-making firm of Lever Brothers
in 1928, "was there was no likelihood of a shortage of whales
generally."[32]
   Outside the boardrooms, however, a few voices were raised

in alarm. "No one unconnected financially with the business of whaling can possibly approve of the present methods of unrestricted slaughter," wrote one observer in 1932.[33] And indeed, he was right. The leading whaling nations were growing concerned that unregulated hunting might lead to the annihilation of the animals, and the end of a very profitable industry. As a result, the 1930s saw the beginning of an international movement to bring the hunt under control before it self-destructed.

WARDS OF THE WORLD

*The world's whale stocks are a truly international
resource in that they belong to no single nation, nor to a
group of nations, but rather they are the wards of the
entire world.*[1]

—American Secretary of State Dean Acheson, 1946

Dean Acheson addressed these remarks to an international
whaling conference convened in Washington, D.C. Anxious to
bring the hunt under control before the whaling fleets plun-
dered the oceans of their last whales, fifteen whaling nations
came together at the end of 1946 to ensure the rational
development of the industry. Following two weeks of negotiation, the conference issued its prescription for saving the
whaling industry from imminent collapse—as conference delegates put it, "to provide for the proper conservation of whale
stocks and thus make possible the orderly development of the
whaling industry."[2]

Chief among the measures adopted by the convention was
the creation of an International Whaling Commission (IWC).
Armed with authority to police the fleets, the Commission was
supposed to herald a new era in whaling, the transformation of a
rapacious hunt into an orderly harvest. At any rate, that was the
convention's hope. Only time would tell if the IWC proved any
more effective than several earlier measures that whalers had
managed either to evade or to ignore.

I

In the early years of the century, when whalers based their activities at shore stations, regulation of the hunt remained in the hands of the individual nations involved. South Georgia, for instance, belonged to the British-owned Falkland Island Dependencies. To avoid overfishing, British authorities issued licences to a small number of whaling companies and restricted their operations by limiting the number of catchers, banning the killing of female whales with calves, and requiring complete processing of the carcasses. In the North Atlantic, whaling was hedged around with similar restrictions in an attempt to protect local fish stocks.

No country went further than Norway to regulate its whalers. First of all, in 1904 it banned whaling off Finnmark. Then, as it became evident that deep-sea whalers in the Antarctic had to be reined in, Norway passed the first law to control whaling on the high seas. The Norwegian Whaling Act (1929) was a landmark piece of legislation. It provided the inspiration for many regulations that would be used again and again in the years ahead. The Act prohibited the killing of right whales, all calves of any species, and any female in the company of a calf. For all species it set minimum lengths below which the animals could not be killed. It required factory ships to carry inspectors and to keep accurate catch records. It encouraged the full use of whale carcasses. It formed a committee charged with collecting and publishing annual whaling statistics. And perhaps most important in the long run, it created a role for scientists in the formulation of whaling policy.[3]

No sooner was the Norwegian law in place than disaster struck the whaling industry. During the southern season of 1930–31, whalers killed 37,438 whales, mostly blues.[4] Oil flooded the market, driving down the price at a time when there were few enough buyers anyway. Thousands of tons of oil sat spoiling in warehouses, and several companies agreed not to send out expeditions the following season. It was evident to the whalers that national controls were not enough. Some kind of international regulation was necessary to stabilize the wild price fluctuations afflicting the industry.

II

The League of Nations initiated the first attempt at international control in 1930. Delegates from several nations drew up a set of proposals modelled for the most part on Norway's whaling law. The following year, twenty-six nations adopted the Geneva Convention for the Regulation of Whaling. However, the document did not come into effect immediately. It first had to be ratified by at least eight governments, and this took until early 1935.

By the standards of later controls, the Geneva Convention made few demands on pelagic whalers. It prohibited the killing of right whales, young whales and females with calves; it required that vessels be licensed, and that oil be extracted from all parts of the carcass. It asked whalers to provide statistics on their catches. And it ordered shipowners to pay their gunners according to the size, species and oil yield of their victims, not simply by the number of animals killed.[5] This regulation was supposed to encourage gunners to hunt only mature whales.

The Geneva Convention was a beginning. For the first time the industry was regulating itself at the international level. Beyond that it had little impact. It established no catch limits, set no minimum lengths, protected no species other than the right whale, and included no penalties for breaking the rules. Most importantly, Germany, the Soviet Union and Japan did not sign.

As yet, the Soviets were minor players in the pelagic industry, confining their activities to the Bering Sea. The Japanese and Germans, however, were a different story. In the mid-1930s, both countries harboured dreams of expanding their territories, and whaling had a part to play. Nazi Germany consumed huge quantities of margarine, manufactured using imported whale oil. Anxious to reduce their dependency on foreign sources of edible oils and to stem the outflow of their currency, the Germans decided to go whaling themselves and joined the Antarctic fleet during the season of 1936–37.

By that time the Japanese had already been active for two seasons. Japanese whalers enjoyed the advantage of a monopoly on the market for whale meat. While most Europeans and North Americans scorned the meat because of its odour and oily texture, the Japanese loved it. As for their oil, they sold most of it abroad to raise foreign currency to fund their incursions into Manchuria and China. Given the strategic importance of their whaling fleets, neither Germany nor Japan was interested in signing agreements that would limit their effectiveness. By the end of the 1930s, whalers from these two "outlaw" nations accounted for about 30 percent of the world catch.

During the 1930s whaling companies tried to stiffen the Geneva Convention by making agreements between themselves. But voluntary, bilateral agreements were futile. Whalers from one country would not abide by rules that they saw being regularly ignored by their competitors. As a result, in 1937, delegates from the major whaling nations gathered in London to hammer out a new international convention.

The London Convention was an improvement over the Geneva agreement. However, the whaling nations stopped short of effective regulation. Delegates set minimum lengths for blue and fin whales, but these lengths were still high enough to allow the capture of sexually immature animals. They banned pelagic catching of baleen whales, except in the Antarctic, the North Pacific and from shore stations, where 90 percent of world whaling took place. They established an open season for the Antarctic, December 8 to March 7, that was too long to have any effect. And they refused to limit the number of catchers, a measure that might have at least capped the growing slaughter.

The next year, 1938, another set of negotiations failed to reach agreement on the pressing issue of limiting the number of whales killed. Japan refused to attend, and Germany refused to agree to a quota. The conference did manage to win agreement for two measures: the creation of a whale sanctuary in an area of the Antarctic where almost no whaling took place anyway, and the complete protection of Antarctic humpback whales, whose numbers had fallen so low that as far as whalers were concerned they were already extinct.

The purpose of early whaling regulation was never to end the hunt. Whaling nations wanted to preserve the animals, and limit the fleet, only so that the hunt might continue indefinitely. The record of the 1937–38 season revealed the utter failure of these plans. About 55,000 whales died that year, the largest catch to date.[6] Not only that, but more factory ships and catchers than ever before patrolled the grounds. Incredibly, a decade of negotiation, discussion and regulation had resulted in a doubling of the kill and an intensification of the hunt.

III

The Second World War accomplished what years of negotiation had not: a dramatic decline in the whale hunt. During the conflict, factory ships were pressed into service as oil tankers serving the war effort. Most fell prey to enemy fire. Catcher boats had become patrol vessels, minesweepers and submarine chasers. About one-third of the fleet was lost in combat; many of the rest were too run down or damaged to return to whaling. Meanwhile, very few companies mounted whaling expeditions to the Antarctic. Of the shore stations, only Grytviken remained in business. World production of oil was lower than at any time since the beginning of southern whaling early in the century. During the final season of the war, 1944–45, whalers took only six thousand animals.[7]

In February 1944, while war still raged in Europe and American troops fought their way, island by island, to a showdown with Japan, delegates from seven nations met in London to plan the future of whaling in the post-war world. The delegates knew that the world would emerge from the war hungry for oil. They assumed that the Allies would exclude Germany and Japan from the hunt. They assumed also that after several years of reduced hunting, the world's whale stocks would have made a comeback. As a result, victorious whaling nations believed they had a unique opportunity to take effective measures to control the hunt in their own interests.

The protocol signed at the end of the London meeting reaffirmed the provisions of the pre-war agreements. As well, it introduced for the first time an allowable limit on the number of whales killed each year. The limit, or quota, was expressed in terms of the Blue Whale Unit (BWU), a concept that bedevilled whaling regulation for the next three decades.

Because all whale species do not produce the same quantity of oil, the industry came up with a measure of equivalency so that one whale might be compared to another in terms of oil yield. The basic unit was the amount of oil obtained from a single, average-sized blue whale. This hypothetical "average" blue whale equalled one Blue Whale Unit, which in turn equalled 2 fin whales, 2.5 humpbacks or 6 sei whales. At the 1944 meeting, the annual quota was set at 16,000 BWU. In other words, whalers could kill 16,000 blue whales, 32,000 fins, 40,000 humpbacks, 96,000 sei whales, or some combination of the four. Expeditions sent weekly catch reports to the Bureau of Whaling Statistics in Norway. As the fleet approached the quota, the bureau calculated and broadcast the date when all ships had to cease whaling.

The quota system seemed reasonable, even revolutionary at the time. Birger Bergersen, later chairman of the International Whaling Commission, called it "one of the most significant events in the history of whaling regulation."[8] However, the system was an abysmal failure. Populations of whales continued to decline rapidly toward extinction. Without any reliable scientific research to go on, the quota was based more or less on wishful thinking. For instance, the allowable catch of 16,000 blue whales probably exceeded the number of blues surviving in the Antarctic. Set too high to begin with, the quota never caught up with the reality of the hunt. Over the years the limit came down, but only when the whalers began to find it impossible to fill anyway. And because it did not differentiate between species, the system still allowed certain species to be overkilled.

Conservationists would have preferred quotas that set limits on the number of whales each nation's fleet could kill, or quotas that differentiated between species. But whaling nations could not agree on such measures. Instead, the overall quota was imposed, with results totally at odds with what the whaling nations were trying to accomplish. Rather than rationalizing the hunt, the quota sparked a frantic competition as each factory ship tried to take as large a share of the limit as it could, as fast as possible. The number of catcher vessels doubled in just five seasons. Each year catchers grew larger and faster. In their haste to kill as many whales as possible before the season closed, gunners did not wait to pick out the largest whales, nor did processors take the time necessary to ensure that no part of the carcasses were wasted. As a result, factory owners saw their costs rise and their efficiency decline as whales became increasingly hard to find. Owners and governments recognized the futility of the situation, but they were unable to agree on an acceptable formula for sharing out the quota.

IV

Towards the end of the Second World War, the Norwegians had hoped to re-establish the status quo of the early 1930s when they and the British ruled the whaling grounds. To this end the

Norwegian government passed a "crew law" forbidding Norwegian seamen to take jobs with any foreign whaling expedition unless its country of origin was involved in whaling before the war. Japanese and German whalers were not expected to return to the grounds, which left the Norwegians and the British to share the annual quota.

However, a worldwide shortage of fats at the end of the war forced up the price of oil, encouraging other nations to get into the business. In 1946 a Dutch company sent one factory ship to the Antarctic, manned largely by Norwegians in flagrant disregard of the "crew law." At the same time, the Americans allowed

Modern catcher boat. One of the catchers that supplied whales to the shore stations on Canada's Pacific coast.

the Japanese to return to the whaling grounds, over the complaints of the British and Norwegians. The Japanese had been catching whales along their coastline since at least the sixteenth century. Indeed, they pioneered their own technique for entrapping the giant creatures in nets. At the end of the nineteenth century they began using modern catcher boats, and in the 1930s they joined the pelagic "club" with their own factory ships. If the hunt was going to continue, there was no excuse for excluding the Japanese.

The spread of whaling in the post-war period convinced Western whaling nations that it was time to establish a permanent body to supervise and regulate the hunt. Present methods of regulation were far too cumbersome. Every change required yet another international conference, followed by the lengthy process of ratification by each signatory. In search of a way out of this impasse, the United States government convened a meeting of whaling nations in Washington, D.C., in the fall of 1946. American delegates were ready with a draft proposal, and at the end of two weeks of negotiation, the conference announced the creation of the International Whaling Commission. It took another two years for an adequate number of countries to ratify what the convention had done, so that it was the spring of 1949 before the Commission held its first meeting. From that moment on, it became the centre of all efforts to control the whale hunt.

Looking back on the history of the IWC with the 20/20 vision of hindsight, it is easy to see that it failed in its stated purpose of protecting the whales and securing the future of the whaling industry. The Commission presents a sorry tale of commercial greed, blind self-interest, futile diplomacy and callous disregard for the world's resources. However, it is only fair to note that the Commission was facing an unprecedented challenge. Maritime nations considered freedom of the high seas sacrosanct. There were no precedents for the cooperative management of marine resources that threatened any nation's right to follow its own self-interest. Only with great reluctance did nations think of surrendering their sovereignty to a world body.

Furthermore, the environmental movement, which has become so powerful, was in its infancy following the Second World War. Only isolated voices exerted any pressure on whaling nations to preserve the whales. Cetaceans did not enjoy the special status they do today. There was little scientific research into their habits, and no understanding of the inter-relatedness of species in the ocean habitat. In the past two decades, whales have come to symbolize the movement to save the environment. In 1946 they were considered a renewable resource like any other, put on the planet to satisfy human needs, no different from all the other creatures with whom they shared the world's oceans.

The IWC's major problems were built into its structure. The convention that created it was not alterable, except in the unlikely event that all members agreed. Among other things, the convention did not allow the Commission to change the system of allocating the yearly catch quota, or to restrict the number of factory ships or shore stations. From the beginning, then, the Commission was hamstrung by its own organization.

The actual whaling regulations that were liable to change from year to year were set out in the Commission's annual Schedule. Most of the regulations declared at the inaugural meeting in 1949 were familiar ones borrowed from earlier agreements. Delegates set the quota for the Antarctic hunt at 16,000 BWU; continued to protect gray and right whales, along with humpbacks in certain areas; banned the killing of calves and females with calves; declared a limited open season; and set minimum length limits for different species.

The Schedule was alterable, when changes met with the approval of three-quarters of the members of the Commission. However, any member not agreeing with a decision could simply register an objection and would not be bound by the decision. Since other members were not likely to abide by rules that one of their rivals felt free to ignore, the objection procedure meant in practice that every member of the Commission had veto power. Delegates to the 1946 meeting were not blind to the probable effect of allowing members to object with impunity. But they knew also that without the ability to opt out of certain regulations, some whaling nations would not join the IWC at all.[9] Better to have half-hearted participation than whole-hearted rejection of all controls. This willingness to conciliate members for fear of losing them altogether was a continuing problem for the Commission.

Another problem was the failure to devise an effective policing procedure. The Commission could pass as many regulations as it liked, but it made no difference if members were not obeying them. Whaling nations were supposed to appoint inspectors for each factory ship, but it was widely acknowledged that many inspectors took a relaxed approach to their duties. R. B. Robertson, doctor aboard a British expedition during the 1950–51 season, observed ships whaling before the official opening of the season and gunners routinely harpooning under-sized and protected animals.[10] It was a sign of the impotence of the Commission that it took almost twenty years of debate before members agreed on a system of international inspectors.

Of course, the Commission itself was not the problem; its members were. At any time whaling nations could have given teeth to the regulations. They did not because Commission members were unprepared to jeopardize the short-term prospects of their whaling fleets to ensure the long-term survival of the whales and the industry. "Are we here to serve the whaling industry or to serve the world?" asked an American delegate in 1973.[11] Certainly, for the first decades of the Commission's existence, most of his colleagues would have believed the question was a little naïve.

V

The International Whaling Commission had little enough success trying to control the activities of its member nations. There was nothing it could do about pirate whalers who flew the flags of non-member nations and operated with complete disregard for the future of the industry.

During the 1950s, the most notorious of these pirates was the Greek shipping magnate, Aristotle Onassis. Onassis did not become one of the richest men in the world playing by the rules. He prided himself on being an outsider, a street fighter, the kind of upstart who, when the other tycoons wouldn't let him join their club, bought the clubhouse, usually with someone else's money. Following the Second World War, Onassis launched a fleet of huge supertankers for transporting oil. The search for shipyards to build these giant vessels brought him to Germany, and to whaling.[12]

Under the terms of the 1945 Potsdam Agreement, the Allies refused to allow German vessels back onto the whaling grounds and prohibited German shipyards from building ships larger than 15,000 tons. But, according to his biographer, Peter Evans, Onassis was eager to do business with the Germans and instructed his closest advisor, Costa Gratsos, to find a way. Gratsos noted that while the Potsdam Agreement banned the construction of large vessels, it did not forbid their conversion to different uses. He suggested that Onassis contract with the Germans to convert an old tanker into a whaling factory ship. This allowed Onassis to establish a business relationship with the Germans that he was able to exploit when the restrictions of the Potsdam Agreement ended.

Onassis was no stranger to the world of whaling. His mistress during the 1930s, Ingeborg Dedichen, was the daughter of Ingevald Martin Bryde, a prominent Norwegian shipowner and son of Johan Bryde who built the first modern whaling station at Durban in South Africa. (Bryde's whale is named after him.) Ingeborg grew up in the Bryde mansion in Sandefjord and introduced Onassis to the leading Norwegian shipowners, many of whom were involved in whaling.

This did not stop the Norwegian whaling fraternity from closing ranks when its members learned about Onassis's plans to enter the business after the war. The international quota was small enough as it was; they did not relish sharing it with another expedition. After failing to intimidate the German shipyard into reneging on its agreement with Onassis, the Norwegians tried the direct approach, offering to buy him out. Along with British whaling interests, they offered to give Onassis 4,000 tons of whale oil a year for the next four years if he called off his plans. This bribe would have amounted to over $4 million, double what Onassis's fleet cost him. But he refused to make a deal, and in October 1950, the converted factory ship *Olympic Challenger* set sail from Germany bound for the Antarctic whaling grounds.

Owned by the Olympic Whaling Company, an Onassis company based in Uruguay, the ship was registered in Panama and carried a crew consisting mainly of Germans, many of them former Nazis. The captain of the *Challenger* was Wilhelm Reichert, latterly an officer in the German navy. The gunners were Norwegians, hired in violation of the "crew law." The manager of the operation was Lars Andersen, one of the best gunners in the business but a pariah in his native Norway since his collaboration with the Nazis during the war. Onassis found him in Buenos Aires advising Juan Perón on Argentina's whaling fleet.

Panama was a convenient home port because it had not signed the Washington Convention, a fact Onassis used to justify the violation of almost every regulation then in force. The *Challenger* fleet began hunting sixteen days before the season officially opened, and continued for two weeks after it closed. Onassis's men killed whales that were smaller than the minimum lengths set by the IWC and systematically falsified their catch records in an attempt to disguise their violations. As a result, the expedition achieved a full cargo of 126,522 barrels of oil, which it trans-shipped to a tanker at the end of the 1950–51 Antarctic season before heading for the coast of Peru to hunt sperm whales.

Onassis had not informed his crew of this plan and the men rebelled, demanding overtime pay before they would go back to work. The owner himself arrived on the factory to negotiate with the whalers, bringing along a party of business associates who amused themselves by taking turns firing the harpoon gun. While his friends treated the excursion like a big-game hunt, Onassis tried without success to convince his men to stay at work past May 10. He failed, and the *Challenger* returned home a month earlier than its owner would have liked. Still, the expedition turned a substantial profit. The outbreak of the Korean War had forced up the price of whale oil, and Onassis claimed to have netted $4 million from his first expedition.

Onassis flaunted his new success as a hunter of whales. In the bar aboard his opulent 322-foot private yacht, the *Christina*, the foot rests were made of sperm whale teeth and the stools were covered with whale foreskins. ("Madame, you are now sitting on the largest penis in the world," he is reported to have said to guest Greta Garbo.) For four seasons Onassis thumbed his nose at the Whaling Commission and its regulations, preying on undersized whales and protected species, out of season, in restricted areas. Disaffected crew members on the *Challenger* later reported seeing baby sperm whales harpooned that had not even grown teeth. They also told how the factory sometimes took ten times as many whales of some species as it reported. During one season the *Challenger* went to the opposite extreme, submitting inflated catch figures so that the allowable quota appeared to have been reached early. In accordance with regulations, the other factories stopped whaling, leaving the Onassis expedition to continue the hunt without competition.

Onassis kept a sharp eye on oil prices. In 1953, when the price dropped suddenly, he suspended whaling for a season, using the *Challenger* as a fuel-oil tanker instead. The next year, when the price improved, he was back in the whale business. But 1954 proved to be no ordinary year for the fleet. Earlier, three South American countries, Chile, Peru and Ecuador, had signed an accord that, among other things, extended their territorial limit two hundred miles out to sea. Other maritime nations were outraged, but they were unwilling to sponsor an outright challenge to the new claim. Onassis did not share their reticence. In the fall of 1954, he sent the *Challenger* into the protected zone to take sperm whales. Peru invoked the new accord with relish, and the "War of the Whales" was on.

On November 15, Peruvian marines boarded three of the catcher boats and ordered them into port. The next morning a Peruvian fighter plane appeared in the sky above the *Challenger*. When the captain made a run for the open sea, the airplane raked the deck with machine-gun fire and dropped warning bombs off the bow. The *Challenger* surrendered, along with another catcher, and headed for shore with a destroyer escort. The rest of the fleet escaped to Panama.

Claiming that the *Challenger* had killed somewhere over 2,500 whales inside their territorial limit, the Peruvians imposed a fine of $3 million, payable within five days. Otherwise, ships and cargo would be sold at auction. The incident caused an international sensation, but Onassis was unperturbed. (When asked by a London reporter what he intended to do about his whaling fleet, he answered, "Why, build another one, of course.") It turned out that he had insured the expedition with Lloyd's of London. Using the insurer's money, he paid the fine and the *Challenger* resumed the hunt.

But the fun was going out of being the bad boy of the whaling industry. At the end of 1955, Norway mounted an effective campaign to expose the discreditable practices of Onassis's fleet. The Norwegians published evidence provided by members of the *Challenger*'s own crew that proved beyond a doubt the charges that had been circulating for several years. At least half of all the whales killed by Onassis's fleet, charged the Norwegians, were killed illegally. Onassis decided it was time to get out of whaling. Early in 1956, he arranged to sell the *Challenger* and its fleet of catchers to a Japanese company for $8.5 million.

VI

The Onassis episode showed how powerless the Whaling Commission was to enforce its regulations. Year after year delegates met in solemn negotiations and set limits on the hunt. And year after year whale populations plunged downward. Finally, in 1959, the commission reached a crisis, and threatened to break apart.

For years, scientists had been warning that baleen whales, especially the blue and the fin, faced extinction unless whalers cut back their catch. Time might already have run out for the blue, the scientists warned; the fin could be saved if drastic measures were taken.

But the whaling nations would not listen. Like spoiled children, they bickered over their share of the annual catch. The Netherlands simply refused to admit there was a problem, or to allow any significant reduction of the quota. Meanwhile, the Soviets announced a plan to quadruple the size of their fleet of factory ships. Two of these vessels were 33,000 gross tons, the largest ever built. To go along with the factories, the Soviets launched sixty-seven new catcher boats.[13] This expansion made no economic sense; the Soviets could not hope to kill enough whales to pay for it. Since the fleet was state-run, however, the normal economic rules did not apply. The other whaling nations now faced the possibility that this aggressive Soviet fleet might gobble up nearly the entire quota.

Obviously, the quota system was ineffective. More drastic measures were needed. The most frequently heard suggestions were either to place a limit on the number of factories taking part in the hunt, or to divide the Antarctic quota between the whaling nations. By securing them a share of the annual hunt and ending open competition, the latter measure might make whalers agreeable to more stringent regulation. It was a sign, however, of the fundamental weakness of the Whaling Commission that under the terms of its charter it did not have the power to enforce either of these solutions.

In the autumn of 1958, the five Antarctic pelagic whaling nations–Britain, Norway, Japan, the Netherlands and the USSR–met in London to find a way out of the impasse. Delegates hoped to agree on a formula for sharing the annual quota. But negotiations foundered, chiefly because the Dutch demanded a share far in excess of what the others were willing to give up.[14] With that, Norway, the Netherlands and Japan all announced they would withdraw from the Whaling Commission at its next meeting. If these countries did walk out, what was the point of anyone remaining and continuing to submit to regulations that their rivals were ignoring? The IWC seemed to be on the verge of dissolution.

The threat of mass defections shifted the spotlight away from the future of the whales and onto the future of the Whaling Commission. Whenever that happened, the former usually was sacrificed to the latter, and 1959 was no exception. Anxious to placate the defectors, the IWC took no steps to antagonize them. The Antarctic quota stayed at 15,000 BWU and nothing was done to reduce the catch of any species, no matter how endangered. This satisfied the Japanese, who withdrew their threat to quit, but after several days of backroom bargaining the Dutch walked out, followed by the Norwegians. An exodus of members was forestalled when Norway announced it would ask its whalers to abide by all the Commission's regulations anyway.[15]

These events initiated a shameful period in the history of whaling. Between 1958 and 1962 whaling nations met eleven times to try to reach agreement about the quota system. While evidence mounted that whales faced annihilation, the hunt actually intensified. Failing to reach an accord on shared quotas, the five Antarctic whaling nations set their own limits, with the result that the 1959–60 season saw a BWU quota that was higher than at any time since the war. The following season, unrestricted hunting in the Antarctic resulted in the death of 41,289 whales, the second largest catch in history.[16] The long years of negotiation and regulation had culminated in almost unprecedented slaughter.

Eventually, in 1962, the major whaling nations settled the issue of quotas. Norway and the Netherlands rejoined the IWC, and life at the Commission returned to normal. Scientific advisers reported with increasing alarm on the state of the whales, while the major whaling nations bargained desperately to forestall the inevitable decline of the industry.

For some the end came more quickly than for others. Whaling was an exceedingly costly enterprise. The modern factory fleet was a small navy, complete with spotter planes and sonar to track down its prey. Along with factories and catcher boats, whalers added freezer ships to transport the meat and buoy boats to tow dead whales back to the factory. In the 1960s a giant factory ship cost $16 million to build and each catcher about a half million dollars.[17] The cost of a single expedition–a factory ship and its entourage of catchers–topped $20 million. Owners needed healthy returns to pay for this capital investment. However, the price of oil was declining through the 1950s and early 1960s. At the same time, whales became harder to find and labour costs rose steadily.

Two of the "Big Five" whaling nations avoided some of these problems. Traditional economics did not impede the Soviets. They kept their whaling fleet at sea even though they could have purchased the oil they produced at half the cost on the open market.[18] The Japanese, for their part, profited from a healthy market for whale meat, a product European and American whalers still could not convince their consumers to eat.

The other three pelagic whaling nations could not compete. One by one they dropped out of the business. In 1963 the last British company withdrew, followed the next year by the Dutch. Then, in 1968, the last Norwegian factory ship, *Kosmos IV*, came home from the Antarctic grounds with 292 BWUs worth of product. Such a meagre kill did not support a factory fleet and Norway, the founder of modern whaling, a country that fifty years earlier accounted for 70 percent of the world catch, dropped out of southern pelagic whaling.[19]

That left the Soviets and the Japanese to carry on the pelagic hunt. In the Antarctic few whales remained to hunt. The IWC finally banned the blue whale kill in 1965; the previous season whalers had found only twenty to kill. Since the humpback was already protected, fin and sei whales became the gunners' prime targets. At the end of the 1960s, whalers were so desperate that they added minkes to the list. The smallest of the rorquals, the minke had been ignored by the whalers because it took so many of them to equal a single blue. However, in the Antarctic, beggars could not be choosers, and minkes rose to the top of the whalers' "hit list."

This dwindling Antarctic hunt, reliant on the least productive remains of a once-vast herd, was not sufficient to satisfy the rapacious factory fleets. In the mid-1960s the focus of the industry shifted back to the northern hemisphere, and back to a familiar species, the sperm whale.

After the death of old-style whaling in the nineteenth century, whalers left sperm whales pretty much alone. Following the Second World War, however, the demand for sperm oil increased and so did the numbers of sperm whales killed. By 1960 they accounted for a third of the world catch; seven years later whalers killed more sperm whales than any other species.

Pelagic catching had been banned in the North Atlantic since 1937 so most of this kill took place in the North Pacific Ocean. Japanese factories preyed on the Bering Sea whales, while the Soviets hunted around the Kamchatka Peninsula and the Kuril Islands. In the mid-1960s, when this hunt peaked, the factories were taking almost 30,000 sperm whales a year. The whale that supported the American fleet during the heyday of old-style whaling was once again the mainstay of the industry.

VII

During the first three decades of the International Whaling Commission's existence, almost 1.5 million whales died, more than in any equivalent period of history.[20] Dean Acheson's concept of whales as "wards of the entire world" was in theory a useful one. Whales were a common resource; therefore, all nations could claim a voice in deciding their fate. However, in practice, if everyone owns the whales, then no one in particular does, and each hunter takes as many whales as he can knowing that, if he does not, his rivals will. With a common resource, competitively pursued, there is no incentive for any individual to conserve—the IWC's regulations failed because whaling nations were not really willing to give them a chance.

Perhaps it would not have made any difference if they had. From the point of view of economics, whales are a renewable resource, like wheat or timber, but they renew themselves very slowly. Some experts argue that by the 1930s populations had dropped so low that it was economically more rational to hunt the whales to extinction and reinvest capital elsewhere than it was to try to "farm" them. If this is the case, there never was an economic motive for whalers to preserve the animals.

As the decade of the 1960s drew to a close, the future of whaling and the whales looked grim. The Soviets and the Japanese were the only large-scale pelagic whalers left, but several countries continued to hunt from coastal stations, and despite the familiar warnings from anxious scientists the population of whales continued to drop. It looked as though the whaling industry was destined to commit slow suicide. The only question was whether it would leave any whales behind when it did.

Then, in the 1970s, a group of activists appeared whose concern was not with the future of whaling but with the future of the whales. Rallying world opinion with sophisticated publicity techniques, risking even death to get their message across, these "eco-guerrillas" managed to hijack the IWC and transform it into a forum for the save-the-whale movement. Suddenly, the imperilled whale became a potent symbol for the whole movement to save the planet from ecological disaster. The whalers did not know what had hit them.

Chapter Twelve

MORATORIUM

*But what a grand day it would be in the evolution of
human consciousness if we could collectively decide to
stop killing the whales now, not for economic or political
reasons, but because we were finally able to see through
the barrier of specism that has separated us from all the
other sentient beings with whom we share this planet.*[1]

–Rex Weyler, Greenpeace photographer, 1977

It was a bizarre, David-meets-Goliath, confrontation. On one
side, a huge Soviet factory ship, longer than two football fields,
going about the routine business of killing sperm whales on the
high seas. On the other side, an eighty-foot wooden fishing boat,
the *Phyllis Cormack*, full of hippie protesters flying Buddhist
banners and peace symbols, demanding in the name of the
whales that the killing stop.

The meeting took place in June 1975. Using classified
information pirated out of Norway, the protesters tracked down
the Soviet factory sixty miles out in the Pacific off the coast of
California. Amused at first, the Soviets crowded the railings of
their ship, waving and snapping photographs of the long-haired
intruders who had materialized so suddenly over the horizon.
Who were these crazy-looking people so far from shore in such a
comical vessel? The name Greenpeace meant as little to them as
the Kwakiutl Indian killer-whale crest that decorated the *Cormack*'s sail. Why should it? At the time, Greenpeace was just a
ragtag collection of mystics, ecology freaks and antiwar protesters, underfunded, barely organized, known to the world–if at
all–for their campaigns against nuclear weapons testing.

Greenpeace activists in Zodiacs try to convince a Soviet catcher boat to give up the hunt.

How were the Soviets to know that these self-styled "Warriors of the Rainbow" were the advance guard of an army of protesters who would hold the whalers up to the world as butchers and exterminators, and eventually help to force an end to the hunt?

The good humour evaporated when the Greenpeace vessel began beaming its anti-whaling message at the factory. The Soviets were not amused when the recorded sounds of humpback whales came floating across the water. "Why can't we live in harmony?" serenaded the protesters. The whalers answered with shouted obscenities.

Tiring of this game, one of the catcher boats, the *Vlastny*, pulled away to resume the hunt, a signal to the Greenpeacers that it was time to put their battle plan into action. The fishing boat was much slower than the catcher and could not hope to keep up with it. However, late in the afternoon, the protesters managed to get close to the Russians as they chased a group of sperm whales. At this point the protesters launched their secret weapon—three inflatable rubber Zodiacs powered by outboard motors.

In the lead Zodiac, Bob Hunter, a Vancouver journalist and leader of the expedition, and Paul Watson, a merchant seaman from New Brunswick, raced alongside the Soviet catcher, then veered suddenly right into its path. The plan was to position themselves between the harpoon and the fleeing whales, expecting that the gunner would not fire if he thought that he might hit another human being.

In their tiny rubber craft, Hunter and Watson sped across the water just a few feet ahead of the sharp prow of the catcher boat. If they slowed, they would be ploughed under and drowned, if not chewed up by the catcher's giant propellor. If they got too far in front, they might ride up on the back of one of the terrified whales and capsize. Then, as if their position was not perilous enough, the Zodiac's engine sputtered, and died.

"I remained frozen at the bow...," recalled Hunter later, "as the harpoon ship reared over a wave, flung its whole weight in a mighty swing down into the water scarcely yards away, and was about to grind us under, when the bow wave it was pushing before it seized us, our outboard engine coughed briefly, biting into the wave, and we were lifted and swept lightly aside, the *Vlastny* passing so close at full speed I could have reached out and touched it."[2]

Quickly, Hunter transferred to a second Zodiac and picked up the chase. Once again he found himself pounding along a few yards in front of the onrushing catcher. High above him the harpoon cannon poised for firing. As the Zodiac went down into the trough between two waves and the whales rose to take a breath, the gunner squeezed the trigger. A direct hit. The heavy whale line came down taut in the water like the blade of a guillotine, not five feet from the rubber Zodiac. A muffled explosion told that the harpoon had found its mark.

There was nothing left for the Greenpeacers to do. A female sperm whale was thrashing in its death agony and its furious mate had turned and was charging down on the catcher. It was no place for a featherweight dinghy. Retreating to the *Cormack*, the protesters watched as the enraged bull sperm threw itself at the bow of the catcher. But there would be no revenge. The gunner coolly took aim and fired again. The wounded bull fell back, its blowhole already spouting blood.

Unnerved by their experience with the protesters, the Soviets took on their two victims, then abandoned the whaling and fled away to the south. The *Cormack* could not keep up. Reaching the limit of its fuel supply, it headed for San Francisco. Later the expedition would try to find the Soviet fleet again, without success. There were no more confrontations that summer.

For all the heady bravado of their performance, what actually had Greenpeace accomplished? By their own calculation, their interference had allowed eight whales to escape the harpoon. What were eight whales when thousands were being slaughtered annually? But Greenpeace tacticians did not intend to save the whales one by one. Their intention was to lift the veil of secrecy that surrounded the hunt and let in the harsh light of publicity. The factory fleets were used to carrying on their business in comfortable obscurity. Few people gave any thought to the hunt. Most consumers did not know that their shoe polish contained whale oil, as did their children's crayons, or that the food they gave their dogs contained whale meat.

Greenpeace wanted to change all that. But whereas David killed Goliath with a stone, Greenpeace brought down its Goliath with the power of the electronic image. The camera was Greenpeace's slingshot. Every moment of the confrontation with the Soviets was recorded on film, along with every detail of the hunt. Suddenly, whaling was front-page news. Gruesome images of dead whales and butchered carcasses appeared in living rooms around the world. "With the single act of filming ourselves in front of the harpoon, we had entered the mass consciousness of modern America," Bob Hunter wrote.[3] The whole world was watching.

I

The campaign to stop all whaling received its marching orders three years earlier, in the summer of 1972. The United Nations sponsored its first Conference on the Human Environment, in Stockholm, where delegates from over sixty nations met to discuss ways to stop poisoning the natural world. Among the many recommendations of the conference was a call for a ten-year moratorium on all commercial whaling. The call to arms was issued in the muted language of international diplomacy, but the meaning was clear. No more quotas; no more exceptions; no more tradeoffs. Stop the killing; then begin a decade of intense research to establish the exact condition of the whale populations.

Outside the International Whaling Commission, the moratorium was greeted with enthusiasm. "The steady destruction of the world's whales—mostly by those enterprising capitalists, the Japanese, and those profit-minded socialists, the Russians—is one of the tragedies of this small planet," the *New York Times* editorialized. "The few surviving species of whales, fascinating, complex and largest of the mammals, deserve man's protection and not the continuing violence of his insensate greed."[4]

But whaling nations were horrified at the idea of a moratorium. When the Stockholm plan was introduced at the IWC meeting later in the summer, members rejected it by a vote of six to four, with four abstentions. The USSR and Japan led the opposition; they were joined by Norway, Panama, Iceland and South Africa. A disappointed Russell Train, head of the American delegation, which supported a moratorium, told the press: "People in some of these other countries have a considerable investment in factory ships and whaling equipment, and that's why they want to continue. What they're doing, though, is killing off species after species. And it'll continue unless we can stop it."[5]

Opponents of the moratorium complained that it was too much of a scattergun approach to the problem. Admittedly, certain stocks of whales were at risk, they said, but others were healthy enough to sustain the hunt. The scientific information did not warrant a complete ban on hunting. Surely what was needed was "prudent management" of the whales on a stock-by-stock basis. Also, they argued, a moratorium interfered with the commission's own program of research, for which it needed a steady supply of whales. For years members of the Commission had ignored the advice of their own scientists. Now, they waved the banner of science in defence of the hunt.

If they were not endorsed by the Whaling Commission, the Stockholm resolutions nevertheless had their effect. The threat of a moratorium acted as a solvent on the determination of hard-line whaling nations, and slowly they gave way on other issues. The Blue Whale Unit disappeared, replaced by stock-by-stock quotas. International observers were placed on every factory ship to ensure that whalers followed IWC regulations. The entire Indian Ocean was made a sanctuary where all cetaceans were safe from their pursuers.

The policy that seemed to hold the most promise for the preservation of the whales was the so-called New Management Procedure (NMP), introduced by the Commission in 1974. In theory, each whale stock has an optimum size at which a certain number of animals can be killed without permanently depleting the population. Under the provisions of the NMP, the hunt was confined only to those stocks that had achieved their optimum size. Stocks that overkilling had reduced to less than their optimum size were protected until they recovered. In practice, the new policy turned out to have serious drawbacks. It depended on accurate population figures, but it was no easy matter to count an animal that ranged so widely and for most of the time was invisible to the naked eye. However, the NMP at least attempted to make science the final arbiter of the hunt, not politics.

II

If stablilization of the whale hunt had always been the objective of the Whaling Commission, it certainly was not the aim of environmentalists, for whom saving the whales meant stopping the whaling. In an open letter to the IWC in 1973, a coalition of conservation groups presented their case for a moratorium.

Protected species of whales were still being killed by whalers illegally or accidentally, the conservationists charged. Only a ban would allow severely depleted species to recover. There was no compelling reason to kill whales, anyway. The meat could be replaced by other sources of protein and substitutes were readily available for other whale products. Whales appeared to have a high level of intelligence and social development; therefore it was "scientifically prudent and ethically sound" to learn more about them before any more died. On the matter of ethics, the conservationists believed that the exploding harpoon was a barbaric weapon that caused a long, agonizing death. Morally, economically and scientifically, the hunt was indefensible, they argued. "We do not believe man any longer needs to hunt whales," concluded the letter. "We think that in the name of human dignity, the killing should stop. . . ."[6]

This letter was signed by the Duke of Edinburgh, along with a fair sampling of naturalists and respected scientists. It was the polite voice of protest. At the other end of the conservationist spectrum was an organization like Greenpeace. Founded in Vancouver, British Columbia, in the early 1970s to oppose American nuclear testing in Alaska, Greenpeace was motivated originally less by science than mysticism. One of its early leaders suggested that members gather at Vancouver's English Bay beach and send telepathic messages to the whales, calling them to take sanctuary in the bay. Astrological charts and the *I Ching* guided its first expedition against the Soviet whalers. A group of musicians came along to serenade the whales, possibly to communicate with them. Crew members aboard the *Phyllis Cormack* talked about saving the world from "environmental Stalinism" and carried "planetary passports."

If their methods were unorthodox, their rhetoric inflated, their courage was indisputable. There is no doubt that members of Greenpeace were willing to die for their cause. They proved it again and again on expeditions against the factory fleets by placing themselves in front of the whaler's harpoon. There is also no doubt that their courage was leavened with a canny sense of publicity. The filmed image of unarmed protesters risking their lives to protect the whales was a powerful one. It stimulated a worldwide revulsion against the hunt. To call Greenpeace publicity-hungry was no insult. Members believed in the power of the image to turn public opinion against the whalers. "Mind bombs" is what Bob Hunter called their harassment of the

Walrus Oakenburgh, crewman aboard the Phyllis Cormack's *1975 voyage to harass the Soviet whalers.*

whaling factories. "If crazy stunts were required in order to draw the focus of the cameras that led back into millions and millions of brains," he explained, "then crazy stunts were what we would do. For in the moment of drawing the mass camera's fire, vital new perceptions would pass into the minds out there that we wanted to reach."[7] If you forced people to witness the butchering of whales, you forced them to act to stop it. That was the theory, the hope, of Greenpeace.

The notoriety achieved by Greenpeace sparked a rapid increase in the size of the organization. Soon it commanded a small fleet of protest vessels, tens of thousands of new members and branch offices all over the globe. As it shed its hippie image, operations grew increasingly sophisticated. But results did not come quickly enough for everyone. While Greenpeace embarrassed the major whaling nations with its "crazy stunts," illegal whaling operations flourished out of sight of public attention, completely indifferent to the arguments of morality or conservation. Paul Watson decided that for these whalers, more drastic measures were necessary.

Watson was a leading member of Greenpeace's early campaigns against the Soviet whalers and the seal hunt off the coast of Newfoundland. An experienced sailor with a history of committed protest behind him—Watson had sneaked through a cordon of U.S. National Guards in 1973 to join the besieged Sioux Indians at Wounded Knee—he chafed at the growing respectability of Greenpeace, and at its resolute committment to nonviolent protest.

In 1977 his fellow protesters, unhappy at the tactics he was using to frustrate the seal hunters, drummed Watson out of Greenpeace. Within a few months he founded his own group, Earthforce, conceived as a kind of "hit squad" for the environmental movement. With the backing of Cleveland Amory's New York-based Fund for Animals, Watson bought a nineteen-year-old deep-water trawler, christened it *Sea Shepherd*, and embarked on a search-and-destroy mission. His target was the most notorious pirate whaler since Aristotle Onassis had retired from the business.

The *Sierra* began life in 1960 as a Dutch catcher boat, patrolling the Antarctic whaling grounds as a member of a factory fleet. When the Dutch gave up pelagic whaling, new owners added processing equipment, a freezer plant and a stern

The original crew of the Phyllis Cormack: top row from left to right, Bob Hunter, Patrick Moore, Bob Cummings, Ben Metcalfe, Dave Birmingham; bottom row, from left to right, Richard Fineberg, Lyle Thurston, Jim Bohlen, Terry Simmons, Bill Darnell and Captain John Cormack.

slipway and converted the 683-ton vessel into a combined catcher-factory. Renamed the *Run* and registered in the Bahamas, it operated off the west coast of Africa where it preyed on whales of any species and any size. In 1972 the owners of the *Run* went bankrupt, chiefly because of a $10,000 fine for illegal whaling. A short time later the ship ended up in the hands of a group of South African businessmen. Given a new name, *Sierra*, and Somalian registration, the vessel continued to kill indiscriminately, chiefly for meat that it supplied on contract to Japan.

In the mid-1970s the Taiyo Fishing Company became at least part-owner of the *Sierra*. Taiyo was a huge multinational corporation, one of Japan's largest companies, with extensive interests in whaling. It built whaling vessels, transported meat in its huge refrigerator ships, owned factories that processed meat and oil and controlled dummy companies like the one that owned the *Sierra*. Representatives of the Taiyo company served on the Japanese delegation to the IWC, but Japan refused to admit that it was in any way involved in whaling by countries that did not belong to the IWC.

By 1979 the Japanese could no longer deny their involvement with pirate whaling. Under pressure from the U.S., which threatened to impose trading sanctions if the Japanese industry did not cut its ties with the pirates, Japan agreed to ban the import of whale products from non-IWC countries. Whether or not this decision would have put the *Sierra* out of business is a moot point. As it turned out, Paul Watson got to the pirate first.

On July 15, somewhere in the Atlantic east of the Azores, the *Sea Shepherd* found its prey. The *Sierra* was heading toward Portugal with a load of whale meat. Pulling in behind it, Watson followed the whaler all night until the next morning they arrived at the port of Leixoes. Here Watson miscalculated. Assuming that the *Sierra* would follow, he took the *Sea Shepherd* into the harbour. Instead, the pirate remained outside, preparing to continue its journey. And port authorities refused to give the *Sea Shepherd* permission to leave again.

Watson was frustrated. If he lost the *Sierra* now, he might never find it again. He assembled the crew and revealed his plan. He was going to make a run for it, sneak out of the harbour and ram the *Sierra*. All but two of the crew left the ship. What Watson proposed was illegal and dangerous. They had no desire to end up in a Portuguese jail.

With his two remaining crewmen, Watson somehow managed

to manoeuvre the *Sea Shepherd* away from the dock. Charging out of the harbour, he roared down on the *Sierra*, drifting peacefully offshore. Before the astonished captain of the whaler could get power up, the *Sea Shepherd* smashed into his bow, then swung in a tight circle and rammed him a second time amidships. The hull of the *Sierra* opened like a wound and the whole side of the vessel caved in. Once again the *Sea Shepherd* disengaged and prepared for a knockout blow that would send the pirate to the bottom. But by this time the captain had managed to get his engine going, and the *Sierra* limped away into Leixoes before Watson could cut it off.

His mission accomplished, Watson made a dash for safety. He knew the Portuguese would be coming after him. Not only had he attacked another vessel, he had left port without permission or a pilot. Still, there was a chance of escape if he could reach Spanish waters to the north. The *Sea Shepherd* raced northward. But it was no match for the Portuguese navy. A few hours later a destroyer was herding the protest vessel back to Leixoes.

Watson was free to leave the country, but the *Sea Shepherd* was impounded. The Portuguese eventually demanded a fine of $750,000, more than six times what the vessel cost to begin with. Watson could not afford to pay. In December, he and a companion returned to Portugal, slipped aboard their boat and opened the sea cocks. Having struck the first blow in the "eco-war" against the whalers, the *Sea Shepherd* filled with water and settled slowly on the harbour bottom.

The second blow was not long in coming. Less than two months later, on the night of February 5, 1980, two divers in wet suits rowed furtively through Lisbon harbour. As they approached the *Sierra*, which was in the capital for a refit, the two men slipped into the black waters and attached a magnetic explosive to the iron hull. By the time an explosion rocked the harbour early the next morning, the underwater bombers were already across the border in Spain, denied the satisfaction of watching the pirate whaler wallow and sink.

Authorities naturally suspected that Paul Watson had something to do with the attack on the *Sierra*. Much as he approved the action, however, Watson was standing trial in Quebec at the time on charges arising from his anti-sealing escapades. Apparently even he did not know who was responsible.

Less than three months later, the eco-guerrillas struck

again. This time their target was the whaling fleet belonging to Juan José Masso, Spain's last whaler. The day before the fleet was set to leave the port of Marin, just north of the Portuguese border, magnetic mines blew holes in two of the catcher boats. No one was hurt in the explosions and once again the bombers got clean away.

Meanwhile, Greenpeace, too, kept up its harassment of the whalers. It was not just the pirates. Several countries, like Spain, which joined the IWC in 1979, permitted whaling and turned a blind eye when its whalers killed protected species or violated other IWC regulations. Chile, Peru, Iceland, Taiwan, South Korea and Norway all allowed shore stations on their coasts.

Greenpeace set itself up as unofficial policeman for the Whaling Commission. In 1979, when Japan agreed to stop importing whale meat from countries that did not belong to the IWC, most of these countries joined the organization. This did not stop their neglect of Commission rules, however. Nor did it really stop Japan's import of whale meat. For instance, meat from Taiwan, which was not a member of the IWC, was funnelled into Japan via South Korea, which was.

Working undercover, Greenpeace revealed the Taiwanese

Soviet Factory Ship with catcher boat alongside.

whale-meat scam, and the illegal sale by Japan of a small factory whaler to the Chileans. At sea, the Greenpeace flagship *Rainbow Warrior*, launched in the summer of 1978, picketed Icelandic and Spanish whaling expeditions, using Zodiacs to protect the animals from the deadly harpoons. Icelanders responded by calling out the navy, and gunboats twice forced the Greenpeace vessel to abandon the confrontation.

The Spanish also came after the *Rainbow Warrior* in force. In June 1980 the Spanish navy intercepted the protest vessel while it was disrupting whalers in the Atlantic, came on board and arrested the crew. The warships escorted the *Warrior* to the port of El Ferrol where it was impounded pending payment of a $142,000 fine.

To ensure that the captive vessel did not escape, Spanish authorities posted an armed guard and removed part of the engine. For five months the *Warrior* sat at the dock at El Ferrol. Greenpeace would not pay the fine; the Spanish would not relent. As this waiting game was going on, however, crew members were secretly replacing the missing engine part and preparing the *Warrior* for sea. On a dark night in November, the vessel slipped its moorings and crept down the five-mile channel leading to the open sea. Then it sped away northward to safety in the Channel Islands.

III

Stunts like the flight of the *Rainbow Warrior* and the sinking of the *Sierra* made great publicity for the anti-whaling movement. At times it seemed to have a monopoly on the good guys. Maybe not everyone agreed with Greenpeace founder Bob Hunter, who called whales "a nation of armless Buddhas" and the whalers "carnivorous Nazis."[8] But public opinion could see little reason for the hunt to continue at all, and no excuse for whalers to endanger the existence of any whale species. As long as the choice was between the whales and the mighty engine of industrialism, the choice was clear: Save the whales.

But not all the world's whalers were commercial hunters using high-tech weaponry to track and kill their prey. In the Azores, for example, men still chase sperm whales the old-fashioned way, using rowboats and hand-held harpoons in a hunt that dates back generations. They kill perhaps 130 animals annually. In Siberia, a single modern catcher boat takes between

Alaskan whalers. An Eskimo crew heading out for the whale hunt, ca. 1900.

150 and 200 gray whales each year under special dispensation from the IWC and delivers the carcasses to villages for the coastal aboriginals to eat.

These hunts are still tolerated because the number of animals killed is small and because sperm and gray whales are not endangered species. It was a different matter in Alaska where, at the end of the 1970s, another traditional hunt, long ignored by the rest of the world, became the focus of a bitter dispute between conservationists and native groups.

Long before the arrival of the first whaleship in the Bering Sea, Alaskan Eskimos were hunting the bowhead whale that passed along the coast on its annual migration. As the ice began to disintegrate each spring, leads of water opened close to shore. The Eskimos waited at the ice edge, and when a whale surfaced they attacked it with harpoons from their skin boats. The hunt was an important source of food for them, and an important source of social status. Ceremony and religious mystery surrounded the chase. With the coming of the Yankee whalers to the western Arctic, the hunt was commercialized. Natives sold their whalebone for southern food and supplies or worked as hired hands for the outsiders. When the market for bone collapsed prior to the First World War and American whalers withdrew from the Arctic, the Eskimos returned to their traditional hunt, armed with the darting guns and exploding harpoons the Americans had taught them to use.

In 1931 the bowhead, its numbers sadly depleted by the American whalers, became a protected species. But Alaskan Eskimos, who killed an average of fewer than twenty animals a year, were allowed to continue their hunt. Then, in the 1970s, the number of Eskimos joining the hunt climbed steeply, as did the number of whales killed. In 1977, for example, the Eskimos killed twenty-eight animals. More important, they struck and wounded forty-four more bowhead that then escaped.[9] These animals were not included in the catch figures, but most probably died from their wounds.

Conservationists sounded the alarm. The International Whaling Commission repeatedly warned the American government to do something to rein in the aboriginal hunt. Frustrated at the inaction of the Americans, concerned that the bowhead might be facing extinction, the IWC at its 1977 meeting banned the Alaskan hunt.

The announcement of the ban hit the Eskimos like a

thunderbolt. They were not commercial whalers. The hunt was crucial for their subsistence, for their very survival as an independent people. They had not slaughtered the vast herds until only a few thousand were left. Why, they said, should they have to pay for the white man's negligence and greed? They lived on the land, knew its harsh realities, and were contemptuous of the southern, big-city environmentalists who now claimed to be saving the world. As far as the Eskimos were concerned, the whaling ban was just another assault on their culture by an indifferent southern culture, especially coming as it did virtually without warning. To sum up their anger, they took to wearing T-shirts that said "Save a whale—eat an Eskimo."

Preoccupied with their own agenda, the anti-whaling lobby spared little sympathy for the Eskimos. An exception was the environmental group Friends of the Earth, which sympathized with the Eskimos and sought a solution that took their needs into account. More important to most conservationists, however, was the threat to the bowhead. They asked that, if the Eskimos were allowed to kill off the last remaining whales, wouldn't that imperil their culture more than a ban on the hunt? One journalist grimly described the Eskimo position: "It's a suicide pact. An endangered culture finishing off an endangered species."[10]

At the IWC, the American delegation was caught in a bind. The Eskimos wanted their government to object formally to the decision to ban the bowhead hunt. But an objection would seriously compromise the American position at the Commission. The United States was seen as the leading antagonist of the whaling nations, and a strong supporter of the moratorium. To come out against a Commission decision would result in a serious loss of credibility. The delegation decided not to object.

The issue now moved to the courts, and the backrooms. In the courts, the Alaskan Eskimo Whaling Commission petitioned to have the American State Department order the IWC delegation to file an objection. After meeting with success in a lower court, the Eskimos' petition was eventually denied. Meanwhile, American delegates were hammering out a compromise with other members of the Whaling Commission. At a special meeting in December 1977, the Commission relented. During the next season, Alaskan Eskimos would be allowed to kill a total of twelve whales, or to strike eighteen, whichever came first. At the same time the Americans promised to come up with an accurate count of the number of bowhead surviving in the western Arctic.

The compromise pleased neither side in the dispute. The Eskimos, facing a 50 percent reduction in their hunt, felt betrayed. So did many conservationists. In order to win acceptance for a quota, the Americans had softened their opposition to the activities of the major whaling nations. "The U.S. must not compromise its principles," a coalition of anti-whaling groups warned President Jimmy Carter, "by trading off hundreds and thousands of whales to the commercial whaling nations for a handful of bowhead whales." But this is exactly what the Americans at the IWC seemed to have done.

The bowhead controversy bedevilled the Commission for the next few years. The quota was raised slightly in 1978, then reduced slightly in 1980, satisfying no one. Finally, at the Whaling Commission meeting in 1980, the U.S. won agreement to a three-year quota. While this made little change in the actual number of bowhead killed, it did free the Americans from the annual bargaining sessions that were doing so much to destroy their country's effectiveness at the Commission. During the next two years, when significant conservation measures came to a vote at the IWC, including the moratorium, whaling nations could not blackmail the Americans into softening their position.

Alaskan Eskimos continue to hunt the bowhead under a quota system approved by the IWC. Over the last decade hunters have struck an annual average of almost twenty-five whales. The most recent figures indicate that the bowhead population in the western Arctic numbers about 7,800 individuals and is not endangered by the activities of native hunters.[11] For the time being, therefore, it looks as though the traditional Eskimo hunt will survive.

IV

Australia was the last English-speaking country in the world to permit commercial whaling. The British stopped in 1963; so did New Zealanders. Americans and Canadians stopped in 1972, South Africans in 1976. This distinction became increasingly embarrassing for the Australians. In 1977 the IWC met at Canberra. Conservationists from around the world gathered to stage anti-whaling protests. A forty-foot, white plastic whale was inflated in the hallway of the hotel where delegates were meeting and had to be removed by stabbing it with kitchen knives. Once again, the whole world was watching.

A public opinion poll showed that 66 percent of Australians opposed whaling. Heading into an election in December, Prime Minister Malcolm Fraser announced that he was going to appoint an inquiry to decide whether Australia should remain in the whaling business. It would be the first time any country had held a public investigation of whaling.

Australia's only whaler, the Cheynes Beach Whaling Company, operated out of a shore station on the southwest tip of the country. This station was all that was left of a once-active industry that during the 1950s decimated the herds of humpbacks that migrated along both sides of the continent. In 1963 the humpback fishery collapsed. Only 87 animals were killed. Stocks along the west coast of Australia had declined from about 15,000 whales to just 800. On the east coast the story was the same. All but one of the shore stations closed. The Cheynes Beach Company survived only by switching to sperm whales.[12]

The inquiry came too late to save the humpbacks. Nevertheless, it had an important impact on the international debate. Inquiry chairman Sir Sydney Frost, a noted Australian jurist, heard submissions from over a hundred groups and individuals. During his investigation the Cheynes Beach Company announced that it was shutting down its whaling operation for economic reasons. But by that time, Frost had gone far beyond the fortunes of one company to study the ethical foundations of the industry as a whole. Frost considered not only the economics of whaling, but the nature of the whale. Did whales have a level of intelligence that, in their own environment, was comparable to man's in his? If so, was it ethically acceptable to kill them? Do whales feel pain? If so, are not the methods used to kill them inhumane? Is it acceptable to treat whales like other animals, as a renewable natural resource, or are they, because of their high intelligence, a special case?

Frost concluded that the whale "is one of those animals which for man have a special significance."[13] He was not prepared to say that whales were especially intelligent, only that research suggested they might be, and while the jury was out allowances should be made. "We are confident that, in light of all the facts put to the Inquiry, reasonable Australian citizens would conclude that, now there is no necessity, it is wrong to kill an animal of such special significance as the whale."[14]

Frost's conclusions were a stunning vindication for the conservation movement. He recommended that Australia ban all whaling within two hundred miles of its coastline. Arguing that substitutes existed for all whale products, he recommended that Australia follow the example of Britain and the United States and ban the import of all goods containing whale products. Perhaps most dramatically, he urged the Australian government to take an activist position in trying to achieve a worldwide ban on whaling.

Frost was highly critical of the International Whaling Commission. The commission, he charged, repeatedly ignored its own scientific advice and allowed killing of different species to continue long after their numbers were dangerously depleted. It clung to the Blue Whale Unit long after the quota was discredited. It delayed fixing quotas on individual stocks when such a measure was obviously needed. For almost twenty years it put off placing impartial observers in the factory fleets to ensure that whalers followed the rules. Yet despite listing these shortcomings, Frost still believed that the Commission was the best, the only, vehicle for achieving a ban on whaling. He noted that the Commission had become much more effective since 1972, and he recommended that Australia remain an active member.

V

Ultimately, the IWC vindicated Sydney Frost's faith in it. Frustrated by internal divisions and embarrassed by the criticisms of its many detractors, the Commission made slow progress toward eventual acceptance of a whaling ban. Each year the moratorium came to a vote. Each year it failed to achieve the necessary majority. But each year more anti-whaling nations joined the "club," swelling the forces in favour of the ban.

Finally, at the 1982 meeting of the IWC in Brighton, England, the moratorium passed: after a phasing-out period lasting until 1986, all commercial whaling would end at least until 1991, by which time the moratorium would be reconsidered and perhaps extended. In the meantime, scientists would attempt to come up with reliable censuses for the different whale populations. The vote was twenty-five in favour, seven against, with five abstentions. The opposition included the Soviet Union, Japan, Norway, Peru, Iceland, Brazil and Korea.

Supporters of the moratorium had every right to celebrate.

It was an impressive accomplishment. But it was not the end of whaling. Nations wishing to continue did not give up without a fight. Japan, the Soviet Union, Peru and Norway formally objected to the moratorium, thus serving notice that they planned to ignore it. Japan in particular applied whatever economic pressures it could to prolong the hunt. Several whaling nations used a loophole in the regulations that permitted whaling for purposes of scientific research.

The "eco-guerrillas" played a continuing role in all this manoeuvring. First into action was the *Rainbow Warrior*. Just five months after the moratorium vote, seven Greenpeace activists boarded a whaleship in the Peruvian port of Paita and chained themselves to the harpoon gun. Angry authorities threatened to toss the protesters into jail, but in the end the *Warrior* and its crew were released with a fine. Whereupon the Greenpeace vessel headed north to harass a Soviet whaling station on the coast of Siberia.

Greenpeace gained notoriety for its campaigns against whaling, but its roots were in the anti-war, anti-nuclear movement and it was this involvement that finally did in the *Rainbow Warrior*. In July 1985 the vessel was in New Zealand taking part in a campaign for a nuclear-free Pacific. During the night of July 10, a party of naval commandos from the French Secret Service attached underwater explosives to the hull of the *Warrior* and blew a hole in its side. As the vessel sank, one of the crew, photographer Fernando Pereira, was trapped below deck and drowned.

The disabling of the Greenpeace flagship did not slow down the anti-whaling campaign. Rather it elevated it to a higher level of violence. Greenpeace actions were intended to obtain the maximum publicity. The organization was adamantly non-violent. However, others were less scrupulous. There had already been the bombing of whaling vessels in Europe, and the ramming of the *Sierra* on the high seas. Now Paul Watson and his *Sea Shepherd* group launched a daring assault on the Icelandic whaling industry.

Iceland had agreed to the moratorium, but was still catching whales for scientific purposes. Critics of this loophole claimed it was just an excuse to continue commercial whaling under a different name; there was no need to kill whales to study them. Two of Watson's agents arrived in the tiny island country in the fall of 1986. One dark night during a snowstorm, the pair broke into the whaling station at Hvalfjordur, north of Reykjavik. Laying about them with heavy sledgehammers, they moved through the station smashing machinery, destroying computers and radio equipment, even defacing files and business records. After five hours of destruction, there was hardly a single piece of equipment working in the entire complex. The pair then returned to Reykjavik where they sneaked on board two whaling boats and opened the sea cocks. As the vessels sank to the bottom of the harbour, the saboteurs made their way to the airport and caught a plane out of the country.

On instructions from Paul Watson, no explosives were used in the action and no lives threatened. Nonetheless, Watson's announcement that he was behind the events in Iceland was greeted by a storm of indignation. It was one thing to peaceably harass and disrupt whaling operations with an eye to affecting public opinion. It was something else again to destroy private property and attack a perfectly legal operation, unpopular though that operation might be. Watson saw no need to apologize. In his mind, the whalers were the terrorists. But his methods touched off a spirited debate within the environmental movement.

While environmentalists debated the morality of sabotage, the few remaining whaling nations announced that they had decided to abide by the moratorium. Anti-whaling protests may have had something to do with these decisions. Far more important was the willingness of the United States and the European Economic Community to apply economic sanctions against countries that continued to whale. Bowing to this pressure, Peru stopped commercial whaling in 1984; the Soviet Union and Norway followed suit in 1987.

Japan was the most important holdout. In 1984 the U.S. and the Japanese came to an agreement about the hunt. The Americans, who had become the enforcers of the anti-whaling movement because of their ability to impose economic penalties, agreed not to penalize the Japanese for continuing their hunt past the moratorium deadline. In turn, the Japanese said they would give up whaling in 1988.[15]

However, when 1988 came, the Japanese announced that they were going to send their fleet back to the Antarctic to kill several hundred whales for scientific purposes, as the Whaling Commission permitted. Outraged, the American government accused the Japanese of continuing the commercial hunt under the flimsy pretext of scientific research. What kind of research, it asked, required the murder of its subject?[16]

The Japanese were adamant. They pointed out that the IWC had imposed the moratorium to give time for a proper assessment of whale populations. They contended that the population of minke whales in the Antarctic was large enough to sustain a limited hunt, and that to prove it they needed to kill some to perform studies on the bodies. Admittedly, the meat and oil were sold later to consumers, but they said that was not the purpose of the hunt. Once again the Japanese threatened to pull out of the Whaling Commission, charging the Americans with racism.[17]

In the end, despite the international protest, the Japanese fleet killed 273 minkes in the Ross Sea in 1988 and a similar number in 1989. As well, a limited catch of small whales continues along the coast of Japan. Iceland and Korea still kill a few hundred whales each year, ostensibly for scientific reasons, and aboriginal whaling is permitted in Alaska, Siberia and Greenland. During 1988, slightly under 700 whales were killed worldwide.[18]

VI

Whaling nations like Japan and Iceland have never accepted the moratorium. They stand ready to re-launch their commercial fleets should the International Whaling Commission decide to allow the resumption of whaling in the 1990s. They have a lot of capital tied up in whaling vessels, and there are still valuable products to be obtained from the whale.

Japanese and Icelanders believe they are the victims of a hysterical public relations campaign that has no justification in science. They do not deny that whales were overhunted in the past, and that some species must remain protected. However, as long as some stocks of whales are healthy, the whaling nations argue, then a controlled hunt should be allowed.

Both Iceland and Japan feel victimized by countries like the United States and Great Britain, countries that hunted the great whales for centuries, helping to bring the animals to their present endangered state. Yet the conservationist movement based in these two countries has successfully portrayed Iceland and Japan as the villains, although they say they simply want to continue taking a limited number of animals from stocks that are still healthy. "The pledge to stop whaling in Iceland for the unforeseeable future means, in reality, that one of the new nations that happened to utilize this resource successfully will be punished by those who did not," writes a prominent Icelandic scientist and member of his country's delegation to the IWC. "It would be equivalent to banning the hunting of caribou in the Canadian Arctic because stocks elsewhere had become endangered."[19] And, these two countries say, not only did Western nations slaughter the whales themselves, but they now agree to allow natives in places like Alaska to go on hunting. Whaling is just as crucial to the survival of coastal communities in Japan, say the Japanese, as it is to the Alaskans. Long before the development of factory ships, shore whalers from coastal villages hunted the animals with nets and harpoons. The chase was full of danger, sanctified by religious ritual, and crucial to the economic well-being of the community. It has been estimated that a ban on whaling in Japan would cost about a hundred jobs and bring hardship to the four villages where whaling is still economically important.[20] Why, the Japanese wonder, are Eskimos singled out for special treatment?

Whatever the merits of these arguments, the 1990s will be a turning point in the history of whaling. The International Whaling Commission will be making the most crucial decision in its history: whether to continue the ban on commercial whaling, or to allow a limited resumption of the hunt. Either way, the Commission faces a crisis. If it does not rescind the moratorium, countries like Iceland and Japan may well leave the organization and resume the hunt, unrestricted by any controls other than their own conscience. Should that happen, the Whaling Commission may well collapse. If, on the other hand, the moratorium is lifted, then environmental activists will surely resume their "actions" against the whalers, actions that before the moratorium were becoming increasingly violent.

Regardless of whether commercial whaling is stopped permanently or allowed to resume under strict controls, it will never again achieve the levels of just a half century ago when tens of thousands of animals were slaughtered annually. The large-scale hunt is over, the victim of its own "success." So many whales were killed that the industry was unable to maintain itself at high levels of production. The Whaling Commission and conservation groups administered the death blow to commercial whaling, but it was already seriously wounded by declining profits, rising costs and a depleted stock of animals.

The long-term impact of whaling is still only partially understood. Populations that have been so dramatically reduced cannot always rebound. Whales are no different in this respect from any other animal. They need a minimum number of individuals to perpetuate themselves. It is possible that in some cases the minimum no longer exists and certain stocks of animals may yet simply die out. Of all the great whale species, only the grays have achieved a population size comparable to what it was before the hunt again.

At the opposite end of the spectrum to the gray is the right whale. Mercilessly slaughtered since the beginning of commercial whaling, these animals have been protected since the 1930s. Yet in many places they have not recovered. It is feared, though not proven, that the population of right whales fell so low during the whaling era that inbreeding among the survivors has brought about a decline in their birth rate and reduced the ability of the population ever to recover fully.[21]

It is difficult to describe with any certainty the condition of the world's whale stocks. At the 1989 meeting of the International Whaling Commission, a group of scientists shocked delegates by reporting that the blue whale population in the Antarctic may be only one-tenth the size calculated earlier.[22] Perhaps the whales are not recovering as quickly as scientists expected.

Whatever the long-term impact of whaling, however, in the short run the whalers did not quite exterminate any species of whale. Individual stocks may have been depleted, even killed off in certain areas. The North Atlantic gray whale is extinct, for example, and the same whale may soon disappear in the Northwest Pacific. Other stocks, their numbers reduced below recoverable levels, may yet disappear. But no species of large whale seems to be in immediate danger. The great whales appear to have been saved. After nine centuries of commercial whaling, the killing stopped in time.

Epilogue: The View From Here

It was one of the most ambitious rescue efforts ever mounted in the Arctic. No expense was spared. Three oil companies offered a hand, along with the American air force and coast guard, the Alaskan National Guard, Greenpeace, the government of the Soviet Union and dozens of Eskimo whalers.

The victims were trapped in the ice near Point Barrow, Alaska, apparently fading fast. It was a race against time. As word of their plight reached the outside world, reporters and camera crews hurried north. For two weeks the dramatic story was front-page news. On the evening television in Canada and the United States, heart-breaking photographs of the sufferers took precedence over important national elections. When one of the victims went missing, presumed drowned, the entire world mourned. When the survivors at last escaped to safety, the world let out a collective cheer.

The object of all this attention and concern was a trio of young gray whales. Trying to leave the Beaufort Sea at the beginning of their southern migration in the fall of 1988, the inexperienced youngsters were held up by the ice. Without help they would have drowned. It was a perfectly natural occurrence, one that probably happens regularly. The difference was that this time the world found out about it.

The startling uproar that resulted indicates the degree to which whales have become the sacred cows of the sea. They have become potent symbols for the fight to save our beleaguered environment. After killing the animals for centuries, we are apparently now willing to spare no effort to preserve even two of them. The rescue cost $1.3 million. Its defenders argue that it was money well spent because it raised public awareness about whales and put wildlife on the front pages, however briefly.

But the episode was full of ironies. The Eskimo whalers who worked so hard to free the trapped animals were the same Eskimo whalers who earlier that year killed thirty-five whales with the permission of the International Whaling Commission and the American government. The animals killed by the Eskimos were bowhead, an endangered species, while the gray whale population is now healthy. Soviet icebreakers were crucial to the success of the rescue effort. Yet the two whales that gained their freedom may well mature only to be killed by the Soviets, who are allowed to catch a certain number of grays off the coast of Siberia.

And the biggest irony of all is that the public that watched the unfolding drama of the whale rescue with such an outpouring of concern is the same public that pays no attention whatsoever to far more deadly perils, most of them man-made, still threatening the world's cetaceans. The hunt for the great whales is virtually over. But whales, and their porpoise and dolphin cousins, continue to die by the thousands because of the carelessness and callousness of man.

Every summer thousands of whale watchers come to observe the snow-white beluga whales of Canada's St Lawrence River. Once known to sailors as "sea canaries" because of the unusual chirping sounds they make underwater, belugas are small by whale standards, growing to a maximum length of about fifteen feet and a maximum weight of 1.5 tons. They are found at several other locations around the world, usually in the Arctic. The belugas of the St Lawrence are unique because they live so close to large settlements—which is why they may not live much longer.

In the past few years, autopsies have been carried out on the corpses of beluga whales washed up on shore. They have found that the animals are contaminated by heavy metals and toxic chemicals, including polychlorinated biphenyls (PCBs) and pesticides. Many of the dead whales suffer from perforated ulcers, cancers and abscesses of the lung. In most cases the cause of death is blood poisoning following upon the failure of the animal's immune system, possibly related to the intake of poisonous substances. A prominent Canadian geneticist has called the beluga "the most polluted mammal on earth."[1]

The belugas of the St Lawrence River were hunted by man off and on for three centuries. Relations between the whales and fishermen were never good. The fishermen blamed the whales for eating valuable fish and early this century began pursuing them in boats armed with harpoons and shotguns. Siding with the fishermen, the government of Quebec during the 1930s placed a bounty of fifteen dollars on the animals in a determined effort to reduce the population. It worked. The number of belugas in the river plummeted. The bounties came off in 1939 but hunting continued into the 1960s. Today the number of whales is estimated at no more than five hundred. That is one-tenth of the population a century ago.[2]

Hunters may have reduced the beluga population so low that it cannot recover. High levels of pollution in the river and in the fish that the whales eat are also certainly taking a toll. The construction of hydroelectric dams may have altered water temperatures, driving away the belugas and, perhaps, their prey. Increased ship and boat traffic may have disrupted the animals' feeding and breeding patterns.

If the causes of the catastrophe are not clear, the results are. Not only has the beluga population declined, it is not recovering. The hunt stopped two decades ago. The whales received government protection about ten years ago. Officially, they have been an endangered species since 1983. Yet the decline is continuing.

The example of the St Lawrence beluga highlights the threat still facing the world's whales. The decline of commercial whaling does not guarantee the survival of the whales. The problem is not as simple as that. Whaling must have sparked a revolution in life under the sea, a revolution that we know almost nothing about. If the natural world exists in a balance, each living thing playing its role in the larger ecosystem, then it stands to reason that the removal of one element disrupts the entire system. Since the slaughter of so many baleen whales in the southern ocean, researchers have noticed an increase in populations of seals and other animals that formerly competed for food with the whales. An estimated 150 million tons of krill is now available for other "consumers" that once was devoured by the whales. Can this rich source of protein be harvested by man without jeopardizing the recovery of whale stocks, or interfering with the equilibrium of the southern ocean in some other way? No one knows. "Things are out of kilter in the oceans now," write two whale experts, "and will be so for a long time."[3]

What is clear is that because of the reduced size of many whale populations they are more vulnerable than ever to the steady contamination of their habitat by man. When it comes to endangered species like blue, right and bowhead whales, the death of a single animal is a blow to the survival of the entire population. Oil spills, raw sewage, construction projects, increased boat traffic, industrial pollution, floating garbage and accidents involving nuclear-powered naval vessels daily combine to make the oceans less livable for the animals that inhabit them. Even a seemingly harmless activity like whale-watching might be disruptive.

One urgent cause of death among cetaceans, particularly smaller whale species, porpoises and dolphins, is entanglement in fishing gear. In the North Atlantic, fishermen have depleted the number of capelin and other fish on which large whales feed. As a result, the whales move closer to shore in search of food and become caught in fishing nets.

Entanglement is also a problem for river dolphins in China and Pakistan, harbour porpoises along the coasts of North America, and Dall's porpoises in the eastern Pacific. Hundreds of thousands of these animals die each year when they swim into nets. In the case of the Pacific yellowfin tuna fishery, fishermen actually set their nets around and under schools of spinner and spotted dolphins, knowing that tuna in large numbers like to swim along beneath them. In the seventeen years between 1971 and 1987, purse seiners "accidentally" killed over six million dolphins this way.[4]

Even more destructive is the spreading use of driftnets by fishermen. These nylon screens hang like a curtain in the water and can extend over thirty-five miles. Other animals, including cetaceans, come to feed on the fish caught in the nets and become entangled themselves. Or they simply swim into the nearly invisible nets by accident and die. Pieces of the net that break off drift across the ocean, a constant threat to large marine life.

It will take many more years for the long-term impact of commercial whaling to work itself out; to know if certain populations will recover; to learn if stocks damaged by overhunting may yet, because of their depleted size, fall victim to other hazards. Meanwhile it is important to recognize that while man may have stopped harpooning whales, or almost, we have not stopped killing them. Plundering and poisoning their habitat, we are imperilling the survival of these creatures as surely as the commercial hunters. "Saving the whales" originally meant stopping the hunt. But presumably we did not "save" them only to stand aside while they perished by other means. "Saving the whales" now means saving the oceans, a much more complicated task.

The questions we must face are whether we will stop using fishing techniques that threaten the whales, if we will reduce the harvest of food resources that we share with them or halt industrial projects that threaten their habitat. These are difficult issues, in which a sentimental attachment to these beautiful, mysterious animals may well play a less important role than hard-headed economics. The world must be fed, and "progress" cannot be stopped. Can whales coexist with these imperatives?

It may be the final irony of all that even as the world finally agrees to save the whales they are more seriously threatened than at any time in their existence.

POSTSCRIPT: October, 1990

In July, 1990, the International Whaling Commission held what was supposed to be one of the most important annual meetings in its 45-year history. But the showdown that is brewing over the future of the moratorium on commercial whaling did not materialize. The purpose of the moratorium is to allow scientists the time to gather the information necessary to make reliable estimates of whale populations, and to develop alternative strategies for "managing" the animals. That work was not completed by the 1990 meeting, and the decision about the future of the moratorium was moved back to the 1991 meeting in Iceland.

Recent statements by whaling nations indicate that pressure is building for a resumption of the commercial hunt, at least on a limited basis. In the weeks and months leading up to the 1990 meeting, Norway and Japan both went public with their determination to return to whaling, one way or another. And a third country, Iceland, tried to obtain IWC permission to take 200 minke whales in the mid-Atlantic. That request was turned down, but none of these nations seems prepared to take no for an answer for much longer. Armed with statistics indicating that there are more than 760,000 minke whales in the oceans of the Southern Hemisphere, and smaller, but still substantial, numbers north of the equator, these nations argue that many hundreds of this particular species could be killed annually without threatening the population as a whole.

Despite the ban on commercial whaling, some killing still goes on. During the 1989-1990 season, a total of 707 animals died. The Japanese, with the only factory ship still in operation, killed 330 minkes in the Antarctic under a scientific permit granted by the IWC. Iceland and Norway killed a total of 85 whales, also for "scientific" reasons. The IWC allows the Soviets, the Americans and the Danes to kill a limited number of whales for subsistance for aboriginal people. In 1989-1990, this aboriginal catch came to 292 whales.

The number of whales now being killed for scientific and subsistance reasons is quite small. The number that will die if commercial hunting recommences is substantially larger. The purpose of a renewed commercial hunt would be to obtain whale meat, principally for the Japanese market. However, there are other motivations. Whaling provides jobs and other economic benefits, especially important in a small country such as Iceland. In Norway, fishermen have become convinced that an increase in the whale population is responsible for a reduction in coastal fish stocks. (There is some irony in the fact that a hundred years ago Norwegian fishermen *opposed* whaling because they believed, erroneously as it turned out, that whales chased the fish into coastal waters and made them easier to catch.)

Iceland, Norway and Japan will carry their fight for a suspension of the moratorium to the annual meeting of the IWC in 1991. Given the state of scientific knowledge, it is possible that the commission will allow a limited return to commercial hunting. Even if it does not, these nations seem to be preparing to resume commercial whaling anyway. They need no one's permission to do so, and may calculate that adverse public reaction is less important than the economic benefits. However, a resumption of whaling in the face of IWC opposition will surely bring the effectiveness of the commission into question. The stage is set for a dramatic showdown.

Illustration & Map Credits

Notes

CHAPTER ONE

1. Quoted in Martin Conway, *No Man's Land: A History of Spitsbergen* (Cambridge: Cambridge University Press, 1906), 35.
2. Robert Grenier, "Excavating a 400-year-old Basque galleon," *National Geographic* (July 1985): 58–68.
3. Roger Collins, *The Basques* (London: Basil Blackwell, 1986).
4. This discussion of early Basque whaling relies on Alex Aguilar, "A Review of Old Basque Whaling and Its Effect on the Right Whales of the North Atlantic," in *Right Whales: Past and Present Status*, eds. Robert L. Brownell, Jr., Peter B. Best, and John H. Prescott (Cambridge: Reports of the International Whaling Commission, Special Issue 10, 1986), 191–200; Jean-Pierre Proulx, *Les Basques et la Pêche à la Baleine: XIe au XVIIe siècle* (Ottawa: Parks Canada, 1983), Microfiche Report Series No. 111.

5. This account of Basque whaling in Labrador relies on Alex Aguilar, "A Review"; Selma Barkham, "The Basques," *Canadian Geographic* (February-March 1978): 8–19; Selma Barkham, "A Note on the Strait of Belle Isle during the Period of Basque Contact with Indians and Inuit," *Etudes/Inuit/Studies* 4 (1980): 51–58; Selma Barkham, "The Basque Whaling Establishments in Labrador, 1536–1632–A Summary," *Arctic* (December 1984): 515–19; James Tuck and Robert Grenier, "A 16th-Century Basque Whaling Station in Labrador," *Scientific American* (November 1981): 180–90.

CHAPTER TWO

1. Frederick Martens, "Voyage into Spitzbergen and Greenland," in *A Collection of Documents on Spitzbergen and Greenland*, ed. Adam White (London: Hakluyt Society, 1855), 114.
2. Martin Conway, *No Man's Land* (Cambridge: Cambridge University Press, 1906), 14.
3. This discussion relies mainly on Conway's history of Spitsbergen, *No Man's Land*, and Martin Conway, ed., *Early Dutch and English Voyages to Spitsbergen in the Seventeenth Century* (London: Hakluyt Society, 1894).
4. Conway, *No Man's Land*, 48.
5. Ibid., 113.
6. Ibid., 118.
7. C. R. Markham, ed., *The Voyages of William Baffin, 1612–22* (London: Hakluyt Society, 1881), 54–79.
8. "Journal Kept by Seven Sailers in Greenland," in *A Collection of Voyages and Travels*, ed. John Churchill (London: 1752), 2:349–58.
9. Conway, *No Man's Land*, 179.
10. Jean-Pierre Proulx, *Whaling in the North Atlantic* (Ottawa: Studies in Archaeology, Architecture and History, Parks Canada, 1986), 24.

CHAPTER THREE

1. Hector St John de Crèvecoeur, *Letters from an American Farmer* (London: J. M. Dent and Sons, 1951; first published 1782), 92.
2. Alexander Starbuck, *History of the American Whale-fishery from its Earliest Inception to the Year 1876* (New York: Antiquarian Press, 1964), 1:5n.
3. This account of shore whaling in northeastern North America is from Everett J. Edwards and J. E. Rattray, *"Whale Off!": The Story of American Shore Whaling* (New York: Coward McCann, 1932) and Randall R. Reeves and Edward Mitchell, "The Long Island, New York, Right Whale Fishery: 1650–1924," in *Right Whales: Past and Present Status*, eds. Robert L. Brownell, Jr., Peter B. Best, and John H. Prescott, 201–21.
4. Crèvecoeur, *Letters*, 91.
5. Richard C. Kugler, "The Whale Oil Trade, 1750–1775," in *Seafaring in Colonial Massachusetts* (Boston: Colonial Society of Massachusetts, 1980), 156.
6. Ibid., 155.
7. Ibid., 157.
8. C. W. Ashley, *The Yankee Whaler* (London: George Routledge and Sons, 1938).
9. Harry Morton, *The Whale's Wake* (Honolulu: University of Hawaii Press, 1982), 61.
10. J. Ross Browne, *Etchings of a Whaling Cruise* (London: John Murray, 1846), 63.
11. Ibid., 62.
12. C. de Jong, "The Hunt for the Greenland Whale," in *Special Issue on Historical Whaling Records*, eds. Michael F. Tillman and Gregory P. Donovan (Cambridge: Reports of the International Whaling Commission, Special Issue 5, 1983), 93.
13. Gordon Jackson, *The British Whaling Trade* (London: Adam and Charles Black, 1978), 55.
14. Basil Lubbock, *The Arctic Whalers* (Glasgow: Brown, Son and Ferguson, 1955), 10.
15. William Scoresby, Jr., *An Account of the Arctic Regions* (Newton Abbot: David and Charles, 1969), 2:189.

16. Robert Goodsir, *An Arctic Voyage to Baffin's Bay and Lancaster Sound* (London: Van Voorst, 1850), 127.
17. Lubbock, *Arctic Whalers*, 254.
18. Goodsir, *An Arctic Voyage*, 8.
19. G. B. Goode, ed., *The Fishery and Fishery Industries of the United States* (Washington: Government Printing Office, 1887), Section 5, 2:108.
20. Joseph L. McDevitt, *The House of Rotch: Massachusetts Whaling Merchants, 1734–1828* (New York: Garland Publishing, 1986), 192.
21. Goode, *The Fishery*, 119.
22. Cited in ibid., 120.

CHAPTER FOUR

1. Cited in Edouard Stackpole, *Whales and Destiny* (Boston: University of Massachusetts Press, 1972), 125.
2. J. C. Beaglehole, ed., *The Journals of James Cook on his Voyages of Discovery* (Cambridge: Cambridge University Press, 1961), 2:604n.
3. Stackpole, *Whales and Destiny*, 127.
4. Ibid., 128.
5. Edward Byers, *The Nation of Nantucket: Society and Politics in an Early American Commercial Center, 1660–1820* (Boston: Northeastern University Press, 1987), 229.
6. "Some Account of the Whale Fishery Formerly Carried on from this Harbour," *The Novascotian* (Halifax: June 15, 1825): 194.
7. McDevitt, *The House of Rotch*, 338.
8. Ibid., 352.
9. Cited in Starbuck, *History*, 1:80.
10. Thierry Du Pasquier, *Les Baleiniers français au XIXe siècle (1814–1868)* (Grenoble: Terre et Mer, 1982), 11.
11. Cited in Jackson, *British Whaling Trade*, 108.
12. Cited in William J. Dakin, *Whalemen Adventurers* (Sydney: Angus & Robertson, 1934), 31.
13. Ibid., 43.
14. Cited in L. S. Rickard, *The Whaling Trade in Old New Zealand* (Auckland: Minerva Publishers, 1965), 32.
15. Ibid., 36.

16. Ibid., 37.
17. Stackpole, *Whales and Destiny*, 335–50.
18. Reginald Horsman, "Nantucket's Peace Treaty with England in 1814," *New England Quarterly* 54, no. 2 (June 1981): 180–98.
19. Starbuck, *History*, 99.
20. Jackson, *British Whaling Trade*, 136.
21. The best source on French whaling in the nineteenth century is Du Pasquier, *Les Baleiniers français*.
22. Elmo Paul Hohman, *The American Whaleman* (New York: Longmans, Green and Co., 1928), 45.
23. Charles Wilkes, *Narrative of the United States' Exploring Expedition, 1838–42* (Philadelphia: Lea and Blanchard, 1845), 5:484.
24. Dakin, *Whaleman Adventurers*, 25–26.
25. Chase's narrative is reprinted in Thomas F. Heffernan, *Stove by a Whale: Owen Chase and the Essex* (Middletown, Ct.: Wesleyan University Press, 1981), 22.
26. *National Geographic* (January 1988): 134.
27. Frederick D. Bennett, *Narrative of a Whaling Voyage Round the Globe* (London: Richard Bentley, 1840), 1:339.
28. P. Wray and K. R. Martin, "Historical Whaling Records from the Western Indian Ocean," in. *Historical Whaling Records*, eds. Tillman and Donovan, 214.
29. Starbuck, *History*, 1:97.
30. Morton, *The Whale's Wake*, 113.
31. Hohman, *The American Whaleman*, 194–95.
32. Ibid., 194.
33. Starbuck, *History*, 1:126–27.
34. Hohman, *The American Whaleman*, 191.
35. Ibid., 193.
36. The most thorough account of this episode, which includes Owen Chase's narrative, is in Heffernan, *Stove by a Whale*.

CHAPTER FIVE

1. Hohman, *The American Whaleman*, 6.
2. Ibid., 9.
3. Heffernan, *Stove by a Whale*, 18.
4. Byers, *The Nation of Nantucket*, 94; Daniel Vickers, "The First Whalemen of Nantucket," *William and Mary Quarterly* 40

(October 1983): 568; Elizabeth A. Little, "The Indian Contribution to Along-Shore Whaling at Nantucket," *Nantucket Algonquin Studies* 8 (Nantucket: Nantucket Historical Association, 1981).

5. Cited in Stackpole, *Whales and Destiny*, 324.
6. Briton Cooper Busch, "Cape Verdeans in the American Whaling and Sealing Industry, 1850–90," *The American Neptune* 45, no. 2 (Spring 1985): 104–16.
7. Cited in Robert W. Kenny, "Yankee Whalers at the Bay of Islands," *The American Neptune* (January 1952): 27.
8. Hohman, *The American Whaleman*, 58.
9. Benjamin Doane, *Following the Sea* (Halifax: Nimbus Publishing and the Nova Scotia Museum, 1987), 56.
10. Browne, *Etchings*, 504–05.
11. Hohman, *The American Whaleman*, 16.
12. Ibid., 72.
13. Ibid, 239.
14. Ibid., 102.
15. Browne, *Etchings*, 24.
16. Hohman, *The American Whaleman*, 273.
17. Ibid.
18. Cited in Charles Roberts Anderson, *Melville in the South Seas* (New York: Dover Books, 1966; originally published 1937), 37.
19. Kenny, "Yankee Whalers," 30; Rickard, *The Whaling Trade*, 41.
20. Harold W. Bradley, *The American Frontier in Hawaii, 1789–1843* (San Francisco: Stanford University Press, 1942), 176–80.
21. Rickard, *The Whaling Trade*, 131–32.
22. This account of the *Globe* mutiny is based largely on Edwin P. Hoyt, *The Mutiny on the Globe* (New York: Random House, 1975).
23. Hohman, *The American Whaleman*, 64.
24. Cited in K. R. Howe, *Where the Waves Fall: A New South Sea Islands History from First Settlement to Colonial Rule* (Honolulu: University of Hawaii Press, 1984), 104.
25. Caroline Ralston, *Grass Huts and Warehouses: Pacific Beach Communities in the Nineteenth Century* (Honolulu: University Press of Hawaii, 1978), 24.
26. Bradley, *American Frontier in Hawaii*, 105.

27. Robert Lloyd Webb, *On the Northwest: Commercial Whaling in the Pacific Northwest, 1790–1967* (Vancouver: UBC Press, 1988).

CHAPTER SIX

1. Charles M. Scammon, *The Marine Mammals of the Northwestern Coast of North America* (New York: Dover Publications, 1968; reprint of 1874 edition), 33.
2. Along with Scammon, the sources for this section are David A. Henderson, *Men and Whales at Scammon's Lagoon* (Los Angeles: Dawson's Book Shop, 1972); the excellent articles in Mary Lou Jones, Steven L. Swartz, and Stephen Leatherwood, eds., *The Gray Whale* (New York: Academic Press, 1984); Steven L. Swartz, "Gray Whale Migratory, Social and Breeding Behavior," in *Behavior of Whales in Relation to Management*, ed. G. P. Donovan (Cambridge: Reports of the International Whaling Commission, Special Issue 8, 1986), 207–29; and Randall R. Reeves and Edward Mitchell, "Current Status of the Gray Whale," *Canadian Field-Naturalist* 102 (1988): 369–90.

CHAPTER SEVEN

1. Charles Edward Smith, *From the Deep of the Sea*, ed. C. E. S. Harris (London: Black, 1922), 134.
2. Sherard Osborn, *Stray Leaves from an Arctic Journal* (New York: George P. Putnam, 1852), 55.
3. Robert Goodsir, *An Arctic Voyage*, 42, 51.
4. Smith, *From the Deep of the Sea*, 81.
5. Sir John Leslie, *Narrative of Discovery and Adventure in Polar Seas and Regions* (Edinburgh: Oliver and Boyd, 1835), 452.
6. W. Gillies Ross, *Arctic Whalers, Icy Seas* (Toronto: Irwin Publishing, 1985), 105.
7. W. Gillies Ross and Alan Cooke, "The Drift of the Whaler *Viewforth*," *Polar Record* 14, no. 92 (May 1969): 585.
8. Basil Lubbock, *The Arctic Whalers*, 305–6.

9. Ross and Cooke, "The Drift of the Whaler *Viewforth*," 586.
10. Lubbock, *Arctic Whalers*, 313.
11. Ibid., 318–19.
12. Ross, *Arctic Whalers, Icy Seas*, 85.
13. Ibid., 107.
14. Clive Holland, "William Penny, 1809–92: Arctic Whaling Master," *Polar Record* 15, no. 94 (January 1970): 27.
15. This discussion of William Penny relies on Holland, ibid.
16. The story of the *Diana* is found in Charles Edward Smith, *From the Deep of the Sea.*
17. Ross, *Arctic Whalers, Icy Seas*, 214.
18. Charles H. Stevenson, "Whalebone: Its Production and Utilization," *Bulletin from Johnny Cake Hill* (Winter 1965–66): 8.
19. W. Gillies Ross, *Whaling and Eskimos: Hudson Bay, 1860–1915* (Ottawa: National Museum of Man, 1975), 54.
20. Joseph-Elzéar Bernier, *Cruise of the Arctic, 1906–07* (Ottawa: King's Printer, 1910), 76.
21. Randall R. Reeves and Stephen Leatherwood, "Bowhead Whales," in *Handbook of Marine Mammals*, eds. S. H. Ridgway and R. Harrison (London: Academic Press, 1985) 3:320.

CHAPTER EIGHT

1. Quoted in F. P. Schmitt, C. de Jong and F. H. Winter, *Thomas Welcome Roys: America's Pioneer of Modern Whaling* (Charlottesville: University Press of Virginia, 1980), 20.
2. This discussion of Thomas Roys relies on Schmitt et al., ibid.
3. John R. Bockstoce, *Whales, Ice and Men: The History of Whaling in the Western Arctic* (Seattle: University of Washington Press, 1986), 95.
4. This description of the cruise of the *Shenandoah* relies on Bockstoce, ibid., 103–28.
5. The following description of the Arctic disasters of the 1870s relies on Bockstoce, 143–79; and Everett Allen, *Children of the Light* (Boston: Little Brown, 1973).

6. G. B. Goode, ed., *The Fishery and Fishery Industries of the United States*, Section 5, 2:152.
7. Ibid., 163.
8. John Bockstoce, *Steam Whaling in the Western Arctic* (New Bedford, Mass.: Old Dartmouth Historical Society, 1977), 22–23.
9. Herbert L. Aldrich, *Eight Months in Arctic Alaska and Siberia with the Arctic Whalemen* (New Bedford, Mass.: Reynolds Printing, 1937), 23.
10. Bockstoce, *Steam Whaling*, 37–40.
11. Hartson Bodfish, *Chasing the Bowhead* (Cambridge, Mass.: Harvard University Press, 1936), 40.
12. Royal North West Mounted Police, Annual Report, 1896, Appendix DD, 238.
13. Tom MacInnis, ed., *Klengenberg of the Arctic* (London: Jonathan Cape, 1932), 130.
14. Bockstoce, *Steam Whaling*, 43.
15. Cited in Bockstoce, *Whales, Ice and Men*, 264.
16. Laurier Papers, Public Archives of Canada, p. 78416.
17. Ibid., p. 78417.
18. John R. Bockstoce and Daniel B. Botkin, "The Historical Status and Reduction of the Western Arctic Bowhead Whale (*Balaena mysticetus*) Population by the Pelagic Whaling Industry, 1848–1914," in *Special Issue on Historical Whaling Records* (Cambridge: Reports of the International Whaling Commission, 1983), 139; Howard W. Braham, "Eskimos, Yankees and Bowheads," *Oceanus* 32, no. 1 (Spring 1989): 58.
19. Schmitt et al., *Thomas Welcome Roys*, 63.
20. Ibid., 181–82.

CHAPTER NINE

1. Quoted in Schmitt et al., *Thomas Welcome Roys*, 78.
2. Ibid., 63.
3. Quoted in J. N. Tønnessen and A. O. Johnsen, *The History of Modern Whaling* (Berkeley: University of California Press, 1982), 26. This is the English translation and abridgement of the original four-volume Norwegian edition, published 1959–70.

4. Ibid., 35.
5. Ibid., 208.
6. Jackson, *British Whaling Trade*, 162.
7. F. D. Ommanney, *Lost Leviathan* (London: Hutchinson and Co., 1971), 136.
8. Tønnessen and Johnsen, *History*, 109–10.
9. Ommanney, *Lost Leviathan*, 103.
10. Ibid., 112.
11. Tønnessen and Johnsen, *History*, 105.

CHAPTER TEN

1. Quoted in Tønnessen and Johnsen, *History*, 277.
2. Quoted in Robert Headland, *The Island of South Georgia* (Cambridge: Cambridge University Press, 1984), 26.
3. Ibid., 29.
4. Ibid.
5. Quoted in Edwin Mickleburgh, *Beyond the Frozen Sea* (London: The Bodley Head, 1987), 127.
6. Jackson, *British Whaling Trade*, 276.
7. Ommanney, *Lost Leviathan*, 13.
8. Quoted in Roland Huntford, *Shackleton* (London: Hodder and Stoughton, 1985), 390.
9. R. B. Robertson, *Of Whales and Men* (New York: Alfred Knopf, 1954), 57.
10. Ernest Shackleton, *South* (London: Heinemann, 1919), 283.
11. Jacques-Yves Cousteau and Yves Paccalet, *Whales* (New York: Harry N. Abrams, 1988), 174.
12. *New York Times*, June 20, 1989.
13. George L. Small, *The Blue Whale* (New York: Columbia University Press, 1971), 14.
14. A. J. Villiers, *Whaling in the Frozen South* (Indianapolis: Bobbs-Merrill, 1925), 55.
15. Ibid., 54.
16. Ibid., 57.
17. Ibid., 60.
18. Quoted in Stephen J. Pyne, *The Ice: A Journey to Antarctica* (New York: Ballantine Books, 1988), 142.

19. Villiers, *Whaling*, 149.
20. Ibid., 150.
21. Ibid., 152.
22. Ibid.
23. Tønnessen and Johnsen, *History*, 350.
24. Ibid., 197, 265–66.
25. Ibid., 385.
26. Ibid., 702.
27. Edward D. Mitchell, Randall R. Reeves and Anne Evely, *Bibliography of Whale Killing Techniques* (Cambridge: Reports of the International Whaling Commission, Special Issue 7, 1986), 7.
28. Tønnessen and Johnsen, *History*, 228.
29. Charles Wilson, *The History of Unilever* (London: Cassell and Co., 1954), 2:121.
30. Tønnessen and Johnsen, *History*, 330.
31. Villiers, *Whaling*, 11.
32. Quoted in Jackson, *British Whaling Trade*, 213.
33. Ibid., 216.

CHAPTER ELEVEN

1. Quoted in Tønnessen and Johnsen, *History*, 532.
2. From the preamble to the International Convention for the Regulation of Whaling, 1946, in *International Regulation of Whaling*, ed. Patricia Birnie (New York: Oceana Publications, 1985), 2: Appendix, 690.
3. Tønnessen and Johnsen, *History*, 364.
4. Ibid., 386.
5. Birnie, *International Regulation*, 1:117.
6. Tønnessen and Johnsen, *History*, 457.
7. Ibid., 476.
8. Ibid., 489.
9. Small, *The Blue Whale*, 177.
10. R. B. Robertson, *Of Whales and Men* (New York: Alfred Knopf, 1954), 276.
11. Quoted in Tønnessen and Johnsen, *History*, 676.
12. This account of Onassis's whaling venture is based on Tønnessen and Johnsen, *History*, 534–37, 553–57; Small,

The Blue Whale, 166–71; Peter Evans, *Ari: The Life and Times of Aristotle Onassis* (New York: Summit Books, 1986), 103–04, 134–35, 145–46.

13. Tønnessen and Johnsen, *History*, 583.
14. Small, *The Blue Whale*, 194.
15. Ibid., 195; Birnie, *International Regulation*, 1:247–50.
16. Birnie, *International Regulation*, 1:256.
17. Small, *The Blue Whale*, 80–81.
18. Tønnessen and Johnsen, *History*, 634.
19. Ibid., 631.
20. Ibid., 680.

CHAPTER TWELVE

1. Robert Hunter, *To Save a Whale: The Voyage of Greenpeace* (Vancouver: Douglas and McIntyre, 1978), 118.
2. Robert Hunter, *Warriors of the Rainbow* (New York: Holt, Rinehart and Winston, 1979), 223.
3. Ibid., 231.
4. *New York Times*, August 30, 1972, 36.
5. *New York Times*, June 30, 1972, 5.
6. *The Times* of London, June 25, 1973.
7. Hunter, *Warriors of the Rainbow*, 252.
8. Ibid., 131.
9. Howard W. Braham, "Eskimos, Yankees and Bowheads," *Oceanus* 32, no. 1 (Spring 1989): 61.
10. Quoted in Jeremy Cherfas, *The Hunting of the Whale* (London: The Bodley Head, 1988), 168.
11. Braham, "Eskimos, Yankees and Bowheads," 62.
12. Friends of the Earth, *The Whaling Question: Report of the Inquiry into Whales and Whaling by Sir Sydney Frost* (San Francisco: FOE, 1979), 36.
13. Ibid., 202.
14. Ibid., 204.
15. *New York Times*, February 11, 1988, 1.
16. Ibid.
17. *New York Times*, January 24, 1988, 18.
18. Robert L. Brownell, Jr., Katherine Ralls, and William F. Perrin, "The Plight of the 'Forgotten' Whales," *Oceanus* 32, no. 1 (Spring 1989): 5.

19. Johann Sigurjonsson, "To Icelanders, Whaling Is a Godsend," *Oceanus* 32, no. 1 (Spring 1989): 32.
20. Kathy Glass and Kirsten Englund, "Why the Japanese Are So Stubborn About Whaling," *Oceanus* 32, no. 1 (Spring 1989):47.
21. *New York Times*, November 1, 1988, C4.
22. Ibid., June 20, 1989, 1.

EPILOGUE

1. *New York Times*, January 12, 1988, C1.
2. R. R. Reeves and E. Mitchell, "Catch History and Initial Population of White Whales (*Delphinapterus Leucas*) in the River and Gulf of St Lawrence, Eastern Canada," *Naturaliste canadien* 111:63–121; and Randall R. Reeves and Edward Mitchell, "Hunting Whales in the St Lawrence," *The Beaver* (August-September 1987): 35–40.
3. Stephen Leatherwood and Randall R. Reeves, *The Sierra Club Handbook of Whales and Dolphins* (San Francisco: Sierra Club Books, 1983), 27.
4. Robert J. Hofman, "The Marine Mammal Protection Act: A First of Its Kind Anywhere," *Oceanus* 32, no. 1 (Spring 1989): 23.

Sources

This is in no way intended as a complete bibliography on the subject of whaling. Instead, I have listed some of the books I found most helpful and that a reader wishing to pursue the subject might find most interesting.

INTRODUCTION

The shelves of libraries groan under the weight of books about whales, most of them richly illustrated. As a beginning, I would suggest three. For a short, readable introduction to the science of whales, Everhard J. Slijper, *Whales and Dolphins* (Ann Arbor, Mich.: University of Michigan Press, 1976). For a coffee-table treatment, Jacques-Yves Cousteau and Yves Paccalet, *Whales* (New York: Harry N. Abrams, 1988). For an excellent ready-reference guide to the different species, Stephen Leatherwood and Randall R. Reeves, *The Sierra Club Handbook of Whales and Dolphins* (San Francisco: Sierra Club Books, 1983).

PART ONE

CHAPTER ONE: FIRST WHALERS

As a result of its involvement in the recovery of the *San Juan* at Red Bay, the Canadian government has produced several useful studies of Basque whaling. These include: Jean-Pierre Proulx, *Les Basques et la Pêche à la Baleine: XIe au XVIIe siècle* (Ottawa: Parks Canada, 1983), Microfiche Report Series; Jean-Pierre Proulx, *Whaling in the North Atlantic: From Earliest Times to the Mid-19th Century* (Ottawa: Parks Canada, 1986); Michael M. Barkham, *Aspects of Life Aboard Spanish Basque Ships During the Sixteenth Century, with special reference to Terranova whaling voyages* (Ottawa: Parks Canada, 1981), Microfiche Report Series; as well as *Research Bulletin* 123, 163, 194, 206, 240, 248 and 258 published by Parks Canada between 1980 and 1987.

Selma Barkham has produced several indispensable articles, including "The Basques," *Canadian Geographic* (February-March 1978), 8–19; "Finding Sources of Canadian History in Spain," *Canadian Geographic* (June 1980), 66–73; "A Note on the Strait of Belle Isle during the Period of Basque Contact with Indians and Inuit," *Etudes/Inuit/Studies* 4(1980): 51–58; "The Basque Whaling Establishments in Labrador, 1536–1632–A Summary," *Arctic* (December 1984): 515–19. Another very useful summary of findings based on the Red Bay site is James Tuck and Robert Grenier, "A 16th-Century Basque Whaling Station in Labrador," *Scientific American* (November 1981): 180–90. A discussion of Basque whaling as a whole, in both the Old World and the New, is Alex Aguilar, "A Review of Old Basque Whaling and Its Effect on the Right Whales of the North Atlantic," in *Right Whales: Past and Present Status*, ed. R. L. Brownell, Jr., Peter B. Best and John H. Prescott (Cambridge: Reports of the International Whaling Commission, Special Issue 10, 1986), 191–200.

CHAPTER TWO: WHALE WARS

Several journals and memoirs written by the participants in the Spitsbergen "whale war" have been collected and published.

These include Adam White, ed., *A Collection of Documents on Spitzbergen and Greenland* (London: Hakluyt Society, 1855) and Martin Conway, ed., *Early Dutch and English Voyages to Spitsbergen in the Seventeenth Century* (London: Hakluyt Society, 1894). Robert Fotherby's account is in C. R. Markham, ed., *The Voyages of William Baffin, 1612–22* (London: Hakluyt Society, 1881). The journal kept by the seven men left on Jan Mayen Island is in John Churchill, ed., *A Collection of Voyages and Travels*, vol. 2 (London: 1752).

The most detailed secondary account of Spitsbergen whaling remains Martin Conway, *No Man's Land: A History of Spitsbergen* (Cambridge: Cambridge University Press, 1906). A more recent summary is in Gordon Jackson, *The British Whaling Trade* (London: Adam and Charles Black, 1978).

CHAPTER THREE:
PLOUGHING THE ROUGHER OCEAN

The origins of American shore whaling are well described in Alexander Starbuck, *History of the American Whale-fishery from Its Earliest Inception to the Year 1876*, 2 vols. (New York: Antiquarian Press, 1964); Everett J. Edwards and J. E. Rattray, *"Whale Off!": The Story of American Shore Whaling* (New York: Coward McCann, 1932); Edouard Stackpole, *The Sea-Hunters: The New England Whalemen During Two Centuries, 1635–1835* (New York: J. B. Lippincott, 1953); and Randall R. Reeves and Edward Mitchell, "The Long Island, New York, Right Whale Fishery: 1650–1924," in *Right Whales: Past and Present Status*, ed. Robert L. Brownell, Jr., Peter B. Best and John H. Prescott (Cambridge: Reports of the International Whaling Commission, Special Issue 10, 1986).

The most famous contemporary description of Nantucket is in Hector St John de Crèvecoeur, *Letters from an American Farmer* (London: J. M. Dent and Sons, Everyman's Library Edition, 1951; first published 1782). An excellent academic history of the island is Edward Byers, *The Nation of Nantucket: Society and Politics in an Early American Commercial Center, 1660–1820* (Boston: Northeastern University Press, 1987). A more popular treatment is Edwin P. Hoyt, *Nantucket: The Life of an Island* (Brattleboro, Vt.: The Stephen Greene Press, 1978). The Indian population of Nantucket is described in Daniel Vickers, "The First Whalemen of Nantucket," *William and Mary Quarterly* 40 (October 1983): 560–83; and in two articles by Elizabeth A. Little: "The Indian Contribution to Along-Shore Whaling at Nantucket," *Nantucket Algonquin Studies 8* (Nantucket: Nantucket Historical Association, 1981), and "Nantucket Whaling in the Early 18th Century," *Papers of the 19th Algonquin Conference*, ed. William Cowan (Ottawa: Carleton University Press, 1988), 111–31.

The economics of eighteenth-century whaling are described in Richard C. Kugler, "The Whale Oil Trade, 1750–1775," in *Seafaring in Colonial Massachusetts* (Boston: Colonial Society of Massachusetts, 1980), 153–73.

The classic account of Arctic whaling is William Scoresby, Jr., *An Account of the Arctic Regions*, 2 vols. (Newton Abbot: David and Charles, 1969; first published 1820). More recent works include Basil Lubbock, *The Arctic Whalers* (Glasgow: Brown, Son and Ferguson, 1955); Jackson, *The British Whaling Trade*; and C. de Jong, "The Hunt for the Greenland Whale," in *Special Issue on Historical Whaling Records*, eds. Michael F. Tillman and Gregory P. Donovan (Cambridge: Reports of the International Whaling Commission, Special Issue 5, 1983).

CHAPTER FOUR: WHALING IN PARADISE

The best account of the extension of whaling into the Pacific is Edouard Stackpole, *Whales and Destiny: The Rivalry Between America, France and Britain for Control of the Southern Whale Fishery, 1785–1825* (Boston: University of Massachusetts Press, 1972). Nantucket's post-revolutionary predicament is described in Byers, *The Nation of Nantucket*; Starbuck, *History of the American Whale-fishery*; and Joseph L. McDevitt, *The House of Rotch: Massachusetts Whaling Merchants, 1734–1828* (New York: Garland Publishing Inc., 1986). A good explanation of the island's tactics during the War of 1812 is Reginald Horsman, "Nantucket's Peace Treaty with England in 1814," *New England Quarterly* 54, no. 2 (June 1981): 180–98.

In 1887 the American government published a detailed study of the fishing industry, including whaling. Two parts are relevant: A. Howard Clark, "History and Present Condition of the Fishery" and James Templeman Brown, "The Whalemen, Vessels and Boats, Apparatus, and Methods of the Whale Fishery" in *The Fishery and Fishery Industries of the United States,* ed. G. B. Goode (Washington, D.C.: Government Printing Office, 1887), Section 5, 2:3–293.

Three books deal with whaling in Australia and New Zealand: William J. Dakin, *Whalemen Adventurers* (Sydney: Angus & Robertson, 1934); L. S. Rickard, *The Whaling Trade in Old New Zealand* (Auckland: Minerva Publishers, 1965); and Harry Morton, *The Whale's Wake* (Honolulu: University of Hawaii Press, 1982). For a history of French whaling, see Thierry Du Pasquier, *Les Baleiniers français au XIXe siècle, 1814-1868* (Grenoble: Terre et Mer, 1982). Herman Melville's whaling career is documented in Charles R. Anderson, *Melville in the South Seas* (New York: Columbia University Press, 1937; reprint Dover Books, 1966).

Charles Wilkes's 1845 report from the Pacific was published in five volumes as *Narrative of the United States' Exploring Expedition, 1838-42* (Philadelphia: Lea and Blanchard, 1845). The published accounts of four seamen who went whaling in the Pacific in the nineteenth century are: Frederick D. Bennett, *Narrative of a Whaling Voyage Round the Globe from the year 1833 to 1836,* 2 vols. (London: Richard Bentley, 1840); J. Ross Browne, *Etchings of a Whaling Cruise* (London: John Murray, 1846); Nelson Cole Haley, *Whale Hunt* (London: The Travel Book Club, 1950); and Benjamin Doane, *Following the Sea* (Halifax: Nimbus Publishing and the Nova Scotia Museum, 1987).

The story of the *Essex* has been told many times, most recently in a novel by Henry Carlisle, *The Jonah Man* (New York: Alfred A. Knopf, 1984). The best source, which includes Owen Chase's own memoir, is Thomas F. Heffernan, *Stove by a Whale: Owen Chase and the* Essex (Middletown, Ct.: Wesleyan University Press, 1981).

CHAPTER FIVE: THE AMERICAN WHALEMAN

The most detailed study of labour conditions in the American whaling industry is Elmo Paul Hohman, *The American Whaleman* (New York: Longmans, Green and Co., 1928). The role of Portuguese islanders is described in Briton Cooper Busch, "Cape Verdeans in the American Whaling and Sealing Industry, 1850-90," *American Neptune* 45, no. 2 (Spring 1985): 104–16.

Useful histories of the Pacific islands are: Harold W. Bradley, *The American Frontier in Hawaii, 1789–1843* (San Francisco: Stanford University Press, 1942); Alan Moorehead, *The Fatal Impact* (London: Hamish Hamilton, 1966); Ernest S. Dodge, *New England and the South Seas* (Cambridge, Mass.: Harvard University Press, 1965); K. R. Howe, *Where the Waves Fall: A New South Sea Islands History From First Settlement to Colonial Rule* (Honolulu: University of Hawaii Press, 1984); and Caroline Ralston, *Grass Huts and Warehouses: Pacific Beach Communities in the Nineteenth Century* (Honolulu: University Press of Hawaii, 1978). An interesting article on New Zealand's Bay of Islands is Robert W. Kenny, "Yankee Whalers at the Bay of Islands," *American Neptune* (January 1952): 22–44. The story of the *Globe* mutiny is recounted in Edwin P. Hoyt, *The Mutiny on the* Globe (New York: Random House, 1975).

CHAPTER SIX:
EXTERMINATING THE DEVILFISH

The authoritative book on the gray whale hunt is David A. Henderson, *Men and Whales at Scammon's Lagoon* (Los Angeles: Dawson's Book Shop, 1972). It should be supplemented by the articles in Mary Lou Jones, Steven L. Swartz and Stephen Leatherwood, eds., *The Gray Whale* (New York: Academic Press Inc., 1984). Captain Scammon's own account is in Charles M. Scammon, *The Marine Mammals of the Northwestern Coast of North America* (New York: Dover Publications, 1968; reprint of the 1874 edition). Also useful are Randall R. Reeves and Edward Mitchell, "Current Status of the Gray Whale," *Canadian Field-Naturalist* 102, no. 2 (1988): 369–89; and Steven L. Swartz, "Gray Whale Migratory, Social and Breeding Behavior," in *Behavior of Whales in Relation to Management*, ed. G. P. Donovan (Cambridge: Reports of the International Whaling Commission, Special Issue 8, 1986), 207–29.

CHAPTER SEVEN:
REGIONS OF ETERNAL FROST

There are several published accounts of whaling voyages into the eastern Arctic: William Barron, *Old Whaling Days* (Hull: William

Andrews, 1895); David Duncan, *Arctic Regions: Voyage to Davis Strait* (London: Billings, 1827); Robert A. Goodsir, *An Arctic Voyage to Baffin's Bay and Lancaster Sound* (London: Van Voorst, 1850); Sir John Leslie, *Narrative of Discovery and Adventure in the Polar Seas and Regions* (Edinburgh: Oliver and Boyd, 1835); David Moore Lindsay, *A Voyage to the Arctic in the Whaler "Aurora"* (Boston: Dana Estes, 1911); A. W. Mackintosh, *A Whaling Cruise in the Arctic Regions* (London: Hamilton, Adams & Co., 1884); Albert H. Markham, *A Whaling Cruise to Baffin's Bay and the Gulf of Boothia* (London: S. Low, Marston, Low and Searle, 1875); and A. Barclay Walker, *The Cruise of the "Esquimaux" to Davis Straits and Baffin Bay, 1889* (Liverpool: Liverpool Printing and Stationery, 1909). The mournful story of the *Diana* is told by the ship's surgeon in Charles Edward Smith, *From the Deep of the Sea*, ed. C. E. S. Harris (London: Black, 1922).

A superb history of Davis Strait whaling is W. Gillies Ross, *Arctic Whalers, Icy Seas* (Toronto: Irwin Publishing, 1985). For the Hudson Bay "fishery," see W. Gillies Ross, *Whaling and Eskimos: Hudson Bay, 1860–1915* (Ottawa: National Museum of Man, 1975) and Ross, ed., *An Arctic Whaling Diary: The Journal of Captain George Comer in Hudson Bay, 1903–05* (Toronto: University of Toronto Press, 1984). The events of the 1830s are detailed in A. G. E. Jones, "The Voyage of HMS *Cove*, Captain James Clark Ross, 1835–6," *Polar Record* 5, no. 40 (July 1950): 543–56; and W. Gillies Ross and Alan Cooke, "The Drift of the Whaler *Viewforth* in Davis Strait, 1835–6, from William Elder's Journal," *Polar Record* 14, no. 92 (May 1969): 581–91. A brief biography of William Penny is in Clive Holland, "William Penny, 1809–92: Arctic Whaling Master," *Polar Record* 15, no. 94 (January 1970): 25–43.

An excellent study of the impact of whaling on whale stocks in the eastern Arctic is Edward Mitchell and Randall R. Reeves, "Catch History and Cumulative Catch Estimates of Initial Population Size of Cetaceans in the Eastern Canadian Arctic," *Report of International Whaling Commission* 31 (1981): 645–82.

CHAPTER EIGHT: THE LAST FRONTIER

The definitive history of western Arctic whaling is John R. Bockstoce, *Whales, Ice and Men: The History of Whaling in the Western Arctic*

(Seattle: University of Washington Press, 1986). Also useful is his earlier *Steam Whaling in the Western Arctic* (New Bedford, Mass.: Old Dartmouth Historical Society, 1977). My own *Arctic Chase: A History of Whaling in Canada's North* (St John's, Nfld.: Breakwater Books, 1984) is a less scholarly treatment of the subject.

Two excellent accounts by Arctic whalemen are: Hartson Bodfish, *Chasing the Bowhead* (Cambridge, Mass.: Harvard University Press, 1936); and John Cook, *Pursuing the Whale* (Boston: Houghton Mifflin, 1926).

Captain Roys's career is ably documented in F. P. Schmitt, C. de Jong and F. H. Winter, *Thomas Welcome Roys: America's Pioneer of Modern Whaling* (Charlottesville: University Press of Virginia, 1980). The disasters of the 1870s are the subject of Everett Allen, *Children of the Light* (Boston: Little Brown, 1973).

The history and present condition of the bowhead whale population is summarized in John R. Bockstoce and Daniel B. Botkin, "The Historical Status and Reduction of the Western Arctic Bowhead Whale (*Balaena mysticetus*) Population by the Pelagic Whaling Industry, 1848–1914," in *Special Issue on Historical Whaling Records*, eds. Tillman and Donovan (Cambridge: Reports of the International Whaling Commission, Special Issue 5, 1983), 107–41; and Howard W. Braham, "Eskimos, Yankees and Bowheads," *Oceanus* 32, no. 1 (Spring 1989): 54–62.

PART TWO

CHAPTER NINE: INDUSTRIAL WHALERS

An exhaustive history of modern whaling is J. N. Tønnessen and A. O. Johnsen, *The History of Modern Whaling* (Berkeley: University of California Press, 1982), an English translation and abridgement of the classic study first published in four volumes, 1959–70, in Norwegian. Schmitt et al., *Thomas Welcome Roys: America's Pioneer of Modern Whaling*, is very good on the technical developments that characterized modern whaling. Also good is Jackson, *The British Whaling Trade*. A fine, scholarly history of whaling on the west coast of Canada is Robert L. Webb, *On the Northwest: Commercial Whaling in the Pacific Northwest, 1790–1967* (Vancouver: University of British Columbia Press, 1988). F. D. Ommanney, *Lost Leviathan* (London: Hutchinson and Co., 1971) is a memoir by a British scientist who worked at a modern whaling station.

CHAPTER TEN:
WHALING IN THE SOUTHERN OCEAN

Several books give first-hand accounts of Antarctic whaling: F. D. Ommanney, *Lost Leviathan*; Robert C. Murphy, *Logbook for Grace* (New York: Macmillan Co., 1947); A. G. Bennett, *Whaling in the Antarctic* (London: William Blackwood and Sons, 1931); R. B. Robertson, *Of Whales and Men* (New York: Alfred Knopf, 1954). An exciting description of the pioneer whaling voyage into the Ross Sea is A. J. Villiers, *Whaling in the Frozen South* (Indianapolis; Bobbs-Merrill Co., 1925). South Georgia's history and geography are described in Robert Headland, *The Island of South Georgia* (Cambridge: Cambridge University Press, 1984). George L. Small, *The Blue Whale* (New York: Columbia University Press, 1971) is an award-winning book on the largest animal that has ever lived. An interesting compendium of information about Antarctica is Stephen J. Pyne, *The Ice: A Journey to Antarctica* (New York: Ballantine Books, 1988). Whale-killing technology is described in Tønnessen and Johnsen, *The History of Modern Whaling*, and in Edward D. Mitchell, Randall R. Reeves and Anne Evely, *Bibliography of Whale Killing Techniques* (Cambridge: International Whaling Commission, Special Issue 7, 1986).

CHAPTER ELEVEN: WARDS OF THE WORLD

The history of whaling regulation can be followed in Tønnessen and Johnsen, *The History of Modern Whaling*; George Small, *The Blue Whale*; and Patricia Birnie, ed., *International Regulation of Whaling* (New York: Oceana Publications Inc., 1985). Also useful is Wray Vamplew, "The Evolution of International Whaling Controls," *Maritime History* 2 (1972): 123–39.

Two books by scientists that quantify the hunt are: N. A. Mackintosh, *The Stocks of Whales* (London: Fishing News Books Ltd., 1965) and K. Radway Allen, *Conservation and Management of Whales* (London: Butterworths, 1980).

The International Whaling Commission has published annual reports and special reports since 1950. Also, since 1930, the Committee for Whaling Statistics in Norway has published catch figures in *International Whaling Statistics*.

CHAPTER TWELVE: MORATORIUM

The story of the anti-whaling campaigns of the 1970s is told in two books by Robert Hunter, *To Save a Whale: The Voyages of Greenpeace* (Vancouver: Douglas & McIntyre, 1978) and *Warriors of the Rainbow: A Chronicle of the Greenpeace Movement* (New York: Holt, Rinehart and Winston, 1979); and in Paul Watson, *Sea Shepherd* (New York: W. W. Norton & Co., 1982). The more general history of the moratorium is described in Jeremy Cherfas, *The Hunting of the Whale* (London: The Bodley Head, 1988), and David Day, *The Whale War* (Vancouver: Douglas & McIntyre, 1987).

Sir Sydney Frost's report is republished as *The Whaling Question* (San Francisco: Friends of the Earth, 1979). The controversy over Alaskan Eskimo whaling is summarized in David Boeri, *People of the Ice Whale* (New York: E. P. Dutton, 1983), and G. P. Donovan, ed., *Aboriginal/Subsistence Whaling* (Cambridge: Reports of the International Whaling Commission, Special Issue 4, 1982). Some of the activities of pirate whalers are documented in Craig Van Note, *Outlaw Whalers* (Washington, D.C.: The Whale Protection Fund, 1979); John Frizell, Campbell Plowden and Alan Thornton, *Outlaw Whalers 1980* (San Francisco: Greenpeace, 1980); and *Unregulated Whaling* (San Francisco: Greenpeace, 1983).

The spring 1989 issue of *Oceanus* (32, no. 1) is a collection of articles bringing many of the issues connected with whaling up to date. An article by a respected scientist considering the arguments for allowing the commercial hunt to resume is

John Gulland, "The End of Whaling?" *New Scientist*, no. 1636 (October 29, 1988): 42–47.

EPILOGUE: THE VIEW FROM HERE

A useful place to begin thinking about some of these matters is the special issue of *Oceanus* noted above. Also Jack Terhune, "Marine Survival," *Policy/Options/Politiques* 6, no. 4 (May 1985).

Index